Field Guide to SDWA Regulations

Technical Editors

William C. Lauer
Mark Scharfenaker
John Stubbart

American Water Works Association

Science and Technology

AWWA unites the entire water community by developing and distributing authoritative scientific and technological knowledge. Through its members, AWWA develops industry standards for products and processes that advance public health and safety. AWWA also provides quality improvement programs for water and wastewater utilities.

Copyright © 2006 American Water Works Association.

All rights reserved. No part of this publication may be reproduced or transmitted in any form or by any means, electronic or mechanical, including photocopy, recording, or any information or retrieval system, except in the form of brief excerpts or quotations for review purposes, without the written permission of the publisher.

Disclaimer

This book is provided for informational purposes only, with the understanding that the publisher, editors, and authors are not thereby engaged in rendering engineering or other professional services. The authors, editors, and publisher make no claim as to the accuracy of the book's contents, or their applicability to any particular circumstance. The editors, authors, and publisher accept no liability to any person for the information or advice provided in this book, or for loss or damages incurred by any person as a result of reliance on its contents. The reader is urged to consult with an appropriate licensed professional before taking any action or making any interpretation that is within the realm of a licensed professional practice.

Library of Congress Cataloging-in-Publication Data

Field guide to SDWA regulations / technical editors, William C. Lauer, Mark Scharfenaker, John Stubbart.

p. cm.

Includes bibliographical references and index.

ISBN 1-58321-385-6

1. Drinking water--United States--Purification. 2. Water treatment plants--United States. I. Lauer, Bill, II, Scharfenaker, Mark, III, Stubbart, John M.

TD433.F54 2005

628.1'62021873--dc22 2005054556

Printed in the United States of America.

American Water Works Association

6666 West Quincy Avenue

Denver, CO 80235-3098

ISBN 1-58321-367-8

CONTENTS

Introduction to SDWA Regulations

The principal law governing drinking water safety in the United States is the Safe Drinking Water Act (SDWA), the common name for Title XIV of the US Public Health Service Act. The SDWA works in concert with the Clean Water Act (CWA), which controls the discharge of pollutants into lakes, rivers, and streams. Enacted in 1974, the SDWA authorizes the US Environmental Protection Agency (USEPA) to establish comprehensive national drinking water regulations to ensure drinking water safety. USEPA is authorized to set national drinking water regulations, conduct special studies and research, and oversee implementation of the act.

OVERVIEW

The 1974 SDWA established a cooperative program among local, state, and federal agencies. The act required promulgation of primary drinking water regulations designed to ensure safe drinking water for consumers. These regulations were the first to apply to all public water systems in the United States, covering both chemical and microbial contaminants. The SDWA mandated a major change in the surveillance of drinking water systems by establishing specific roles for federal and state governments and for public water suppliers.

State governments, through their health departments and environmental agencies, received the major responsibility, called primary enforcement responsibility, or *primacy,* for the administration and enforcement of the regulations set by USEPA. The SDWA set a schedule and procedures for developing new drinking water standards, which included health-based goals, known as *maximum contaminant level goals* (MCLGs), and technically achievable standards, known as *maximum contaminant levels* (MCLs). The act also authorized USEPA to establish treatment techniques instead of MCLs when it is not economically or technologically feasible to determine the level of a contaminant.

The MCLs and treatment techniques comprise the National Primary Drinking Water Regulations (NPDWRs) and are federally enforceable. The act also authorized USEPA to establish National Secondary Drinking Water Regulations (NSDWRs), which are nonenforceable standards established to control aesthetic parameters such as taste and odor.

Major Amendments

The SDWA was amended in 1986 and 1996. Major mandates of the 1986 amendments included:

- Setting standards for 83 contaminants by June 1989 and for an additional 25 contaminants every 3 years thereafter
- Designating best available technology (BAT) for each newly regulated contaminant
- Disinfection of all public water supplies and criteria for mandating filtration of surface water supplies
- Monitoring for contaminants that are not regulated to determine if additional regulation is necessary
- Banning lead solders, lead flux, and lead pipe in public water systems
- Implementing new programs for protecting wellheads and sole-source aquifers and preventing contamination of groundwater sources from waste-injection wells

Congress also amended the SDWA in 1988 by adding the Lead Contamination Control Act. The act, among other things, instituted a program to eliminate lead-containing drinking water coolers in schools.

The SDWA amendments of 1996, which govern current program activities, made substantial revisions to the act. The revisions include establishment of a state revolving fund to provide loans to help water systems comply with SDWA regulations and setting aside funds to support program implementation.

The 1996 amendments also added provisions for

- Replacing the 1986 rulemaking schedule with a program that requires USEPA to decide every 5 years whether to regulate at least five contaminants based on their occurrence and risk and the potential for achieving meaningful risk reduction
- Extending to 3 years the previous 18-month deadline for systems to comply with new regulations, with an additional 2 years if necessary for capital improvements
- Establishing specific requirements for regulating arsenic, disinfection by-products, microbial contaminants, and radon
- Requiring community water systems to provide their customers with annual water quality reports, called Consumer Confidence Reports (CCRs)
- Requiring states to assess and delineate source water protection areas for public water systems and to implement a program to certify water system operators

- Encouraging states to implement a program to build the managerial, technical, and financial capacity of water systems to comply with SDWA regulations and
- Establishing a formal program for monitoring unregulated contaminants

Congress most recently amended the SDWA in 2002 to address drinking water safety and security in light of potential terrorism threats. The amendments are part of the Public Health Security and Bioterrorism Preparedness and Response Act, in response to the 2001 terrorist attacks in New York City and Washington, D.C.

The Bioterrorism Act required owners and operators of community water systems serving more than 3,300 people to assess their systems' vulnerability to terrorist attack or other intentional acts aimed at disrupting the supply of safe drinking water and to submit these vulnerability assessments to USEPA. Affected drinking water systems were also required to prepare emergency response plans. The act further required USEPA to research methods to prevent, detect, and respond to the intentional introduction of chemical, biological, or radiological contaminants into community water systems and to provide certain threat-reduction information to drinking water systems.

Drinking Water Regulations

Regulations promulgated by USEPA contain technical details that specify the performance levels necessary to achieve compliance with SDWA provisions. USEPA's Office of Ground Water and Drinking Water (OGWDW), part of the Office of Water, has the primary responsibility for establishing NPDWRs, protecting groundwater, implementing regulations, providing technical support, and enforcing SDWA mandates.

Two types of drinking water standards are promulgated under the drinking water standards program: enforceable primary regulations to control health risks (Table 1-1) and nonenforceable secondary regulations (Table 1-2) to address aesthetic concerns such as taste and odor.

A primary regulation can be an MCL or a treatment technique. An MCL is set based on an MCLG, which is the concentration of a contaminant in drinking water below which there is no known or expected risk to health. The SDWA requires an MCL to be set as close as technically and economically feasible to an MCLG; often the two values are equal. In some cases, the MCLG is not technically and economically achievable, and its respective MCL is less stringent. For known or suspected human carcinogens, the MCLG is usually zero.

When it is impossible or impractical to establish a numeric standard, the law authorizes USEPA to establish a treatment technique and to specify treatment methods that must be used to minimize exposure of the public. This was done for lead and copper regulations.

Another legally enforceable standard was authorized under the 1996 amendments to the SDWA to control disinfectant residuals within distribution systems. It is known as the maximum residual disinfectant level (MRDL), which is also set as close as technically and economically feasible to a health-based MRDL goal. To date, USEPA has set MRDLs for three disinfectants: chlorine, chloramines, and chlorine dioxide.

Table 1-1 List of contaminants and their regulatory standard*

Contaminant	MCLG, *mg/L*†	MCL or TT, *mg/L*	Potential Health Effects From Ingestion of Water	Sources of Contaminants in Drinking Water
Microorganisms				
Cryptosporidium	zero	TT‡	Gastrointestinal illness (e.g., diarrhea, vomiting, cramps).	Human and animal fecal waste.
Giardia lamblia	zero	TT	Gastrointestinal illness (e.g., diarrhea, vomiting, cramps).	Human and animal fecal waste.
Heterotrophic plate count (HPC)	N/A	TT	HPC has no health effects; it is an analytic method used to measure the variety of bacteria that are common in water. The lower the concentration of bacteria in drinking water, the better maintained the water system.	HPC measures a range of bacteria that are naturally present in the environment.
Legionella	zero	TT	Legionnaire's disease, a type of pneumonia.	Found naturally in water; multiplies in heating systems.
Total coliforms (including fecal coliform and *E. coli*)	zero	5.0%§	Not a health threat in itself; is used to indicate whether other potentially harmful bacteria may be present.**	Coliforms are naturally present in the environment, as well as feces; fecal coliforms and *E. coli* only come from human and animal fecal waste.
Turbidity	N/A	TT	Turbidity is a measure of the cloudiness of water. It is used to indicate water quality and filtration effectiveness (e.g., whether disease-causing organisms are present). Higher turbidity levels are often associated with higher levels of disease-causing microorganisms such as viruses, parasites, and some bacteria. These organisms can cause symptoms such as nausea, cramps, diarrhea, and associated headaches.	Soil runoff.

Table continued next page.

Table 1-1 List of contaminants and their regulatory standard* (continued)

Contaminant	MCLG, *mg/L*†	MCL or TT, *mg/L*	Potential Health Effects From Ingestion of Water	Sources of Contaminants in Drinking Water
Viruses (enteric)	zero	TT	Gastrointestinal illness (e.g., diarrhea, vomiting, cramps).	Human and animal fecal waste.
Disinfection by-products				
Bromate	zero	0.010	Increased risk of cancer.	By-product of drinking water disinfection.
Chlorite	0.8	1.0	Anemia; nervous system effects in infants and young children.	By-product of drinking water disinfection.
Haloacetic acids (HAA5)	N/A††	0.060	Increased risk of cancer.	By-product of drinking water disinfection.
Total trihalomethanes (TTHMs)	none‡‡ / N/A	0.10 / 0.080	Liver, kidney, or central nervous system problems; increased risk of cancer.	By-product of drinking water disinfection.
Disinfectants				
Chloramines (as Cl_2)	MRDLG = 4	MRDL = 4.0	Eye/nose irritation; stomach discomfort; anemia.	Water additive used to control microbes.
Chlorine (as Cl_2)	MRDLG = 4	MRDL = 4.0	Eye/nose irritation; stomach discomfort.	Water additive used to control microbes.
Chlorine dioxide (as ClO_2)	MRDLG = 0.8	MRDL = 0.8	Anemia; nervous system effects in infants and young children.	Water additive used to control microbes.
Inorganic Chemicals				
Antimony	0.006	0.006	Increase in blood cholesterol; decrease in blood sugar.	Discharge from petroleum refineries; fire retardants; ceramics; electronics; solder.
Arsenic	0	0.010 as of 1/23/06	Skin damage or problems with circulatory systems; may cause increased risk of cancer.	Erosion of natural deposits; runoff from orchards, glass and electronics production wastes.

Table continued next page.

Table 1-1 List of contaminants and their regulatory standard* (continued)

Contaminant	MCLG, *mg/L*†	MCL or TT, *mg/L*	Potential Health Effects From Ingestion of Water	Sources of Contaminants in Drinking Water
Asbestos (fiber >10 μm)	7 million fibers per liter (MFL)	7 MFL	Increased risk of developing benign intestinal polyps.	Decay of asbestos cement in water mains; erosion of natural deposits.
Barium	2	2	Increase in blood pressure.	Discharge of drilling wastes; discharge from metal refineries; erosion of natural deposits.
Beryllium	0.004	0.004	Intestinal lesions.	Discharge from metal refineries and coal-burning factories; discharge from electrical, aerospace, and defense industries.
Cadmium	0.005	0.005	Kidney damage.	Corrosion of galvanized pipes; erosion of natural deposits; discharge from metal refineries; runoff from waste batteries and paints.
Chromium (total)	0.1	0.1	Allergic dermatitis.	Discharge from steel and pulp mills; erosion of natural deposits.
Copper	1.3	TT§§; action level = 1.3	Short-term exposure: gastrointestinal distress; long-term exposure: liver or kidney damage; people with Wilson's disease should consult their personal physician if the amount of copper in their water exceeds the action level.	Corrosion of household plumbing systems; erosion of natural deposits.
Cyanide (as free cyanide)	0.2	0.2	Nerve damage or thyroid problems.	Discharge from steel/metal factories; discharge from plastics and fertilizer factories.
Fluoride	4.0	4.0	Bone disease (pain and tenderness of the bones); children may get mottled teeth.	Water additive that promotes strong teeth; erosion of natural deposits; discharge from fertilizer and aluminum factories.

Table continued next page.

Table 1-1 List of contaminants and their regulatory standard* (continued)

Contaminant	MCLG, mg/L†	MCL or TT, mg/L	Potential Health Effects From Ingestion of Water	Sources of Contaminants in Drinking Water
Lead	zero	TT; action level = 0.015	Infants and children: delays in physical or mental development; children could show slight deficits in attention span and learning abilities. Adults: kidney problems; high blood pressure.	Corrosion of household plumbing systems; erosion of natural deposits.
Mercury (inorganic)	0.002	0.002	Kidney damage.	Erosion of natural deposits; discharge from refineries and factories; runoff from landfills and croplands.
Nitrate (measured as nitrogen)	10	10	Infants below the age of 6 months who drink water containing nitrate in excess of the MCL could become seriously ill and, if untreated, may die; symptoms include shortness of breath and blue-baby syndrome.	Runoff from fertilizer use; leaching from septic tanks, sewage; erosion of natural deposits.
Nitrite (measured as nitrogen)	1	1	Infants below the age of 6 months who drink water containing nitrite in excess of the MCL could become seriously ill and, if untreated, may die; symptoms include shortness of breath and blue-baby syndrome.	Runoff from fertilizer use; leaching from septic tanks, sewage; erosion of natural deposits.
Selenium	0.05	0.05	Hair or fingernail loss; numbness in fingers or toes; circulatory problems.	Discharge from petroleum refineries; erosion of natural deposits; discharge from mines.
Thallium	0.0005	0.002	Hair loss; changes in blood; kidney, intestinal, or liver problems.	Leaching from ore-processing sites; discharge from electronics, glass, and drug factories.
Organic Chemicals				
Acrylamide	zero	TT***	Nervous system or blood problems; increased risk of cancer.	Added to water during sewage/wastewater treatment.
Alachlor	zero	0.002	Eye, liver, kidney, or spleen problems; anemia; increased risk of cancer.	Runoff from herbicide used on row crops.

Table continued next page.

Table 1-1 List of contaminants and their regulatory standard* (continued)

Contaminant	MCLG, *mg/L*†	MCL or TT, *mg/L*	Potential Health Effects From Ingestion of Water	Sources of Contaminants in Drinking Water
Atrazine	0.003	0.003	Cardiovascular system or reproductive problems.	Runoff from herbicide used on row crops.
Benzene	zero	0.005	Anemia; decrease in blood platelets; increased risk of cancer.	Discharge from factories; leaching from gas storage tanks and landfills.
Benzo(a)pyrene (PAHs)	zero	0.0002	Reproductive difficulties; increased risk of cancer.	Leaching from linings of water storage tanks and distribution lines.
Carbofuran	0.04	0.04	Problems with blood, nervous system, or reproductive system.	Leaching of soil fumigant used on rice and alfalfa.
Carbon tetrachloride	zero	0.005	Liver problems; increased risk of cancer.	Discharge from chemical plants and other industrial activities.
Chlordane	zero	0.002	Liver or nervous system problems; increased risk of cancer.	Residue of banned termiticide.
Chlorobenzene	0.1	0.1	Liver or kidney problems.	Discharge from chemical and agricultural chemical factories.
2,4-D	0.07	0.07	Kidney, liver, or adrenal gland problems.	Runoff from herbicide used on row crops.
Dalapon	0.2	0.2	Minor kidney changes.	Runoff from herbicide used on rights-of-way.
1,2-Dibromo-3-chloropropane (DBCP)	zero	0.0002	Reproductive difficulties; increased risk of cancer.	Runoff/leaching from soil fumigant used on soybeans, cotton, pineapples, and orchards.
o-Dichlorobenzene	0.6	0.6	Liver, kidney, or circulatory system problems.	Discharge from industrial chemical factories.
p-Dichlorobenzene	0.075	0.075	Anemia; liver, kidney, or spleen damage; changes in blood.	Discharge from industrial chemical factories.
1,2-Dichloroethane	zero	0.005	Increased risk of cancer.	Discharge from industrial chemical factories.

Table continued next page.

Table 1-1 List of contaminants and their regulatory standard* (continued)

Contaminant	MCLG, *mg/L*†	MCL or TT, *mg/L*	Potential Health Effects From Ingestion of Water	Sources of Contaminants in Drinking Water
1,1-Dichloroethylene	0.007	0.007	Liver problems.	Discharge from industrial chemical factories.
cis-1,2-Dichloroethylene	0.07	0.07	Liver problems.	Discharge from industrial chemical factories.
trans-1,2-Dichloroethylene	0.1	0.1	Liver problems.	Discharge from industrial chemical factories.
Dichloromethane	zero	0.005	Liver problems; increased risk of cancer.	Discharge from drug and chemical factories.
1,2-Dichloropropane	zero	0.005	Increased risk of cancer.	Discharge from industrial chemical factories.
Di(2-ethylhexyl) adipate	0.4	0.4	Weight loss; liver problems; possible reproductive difficulties.	Discharge from chemical factories.
Di(2-ethylhexyl) phthalate	zero	0.006	Reproductive difficulties; liver problems; increased risk of cancer.	Discharge from rubber and chemical factories.
Dinoseb	0.007	0.007	Reproductive difficulties.	Runoff from herbicide used on soybeans and vegetables.
Dioxin (2,3,7,8-TCDD)	zero	0.00000003	Reproductive difficulties; increased risk of cancer.	Emissions from waste incineration and other combustion; discharge from chemical factories.
Diquat	0.02	0.02	Cataracts.	Runoff from herbicide use.
Endothall	0.1	0.1	Stomach and intestinal problems.	Runoff from herbicide use.
Endrin	0.002	0.002	Liver problems.	Residue of banned insecticide.
Epichlorohydrin	zero	TT	Increased cancer risk; over long periods of time, stomach problems.	Discharge from industrial chemical factories; an impurity of some water treatment chemicals.

Table continued next page.

Table 1-1 List of contaminants and their regulatory standard* (continued)

Contaminant	MCLG, *mg/L*†	MCL or TT, *mg/L*	Potential Health Effects From Ingestion of Water	Sources of Contaminants in Drinking Water
Ethylbenzene	0.7	0.7	Liver or kidney problems.	Discharge from petroleum refineries.
Ethylene dibromide	zero	0.00005	Problems with liver, stomach, reproductive system, or kidneys; increased risk of cancer.	Discharge from petroleum refineries.
Glyphosate	0.7	0.7	Kidney problems; reproductive difficulties.	Runoff from herbicide use.
Heptachlor	zero	0.0004	Liver damage; increased risk of cancer.	Residue of banned termiticide.
Heptachlor epoxide	zero	0.0002	Liver damage; increased risk of cancer.	Breakdown of heptachlor.
Hexachlorobenzene	zero	0.001	Liver or kidney problems; reproductive difficulties; increased risk of cancer.	Discharge from metal refineries and agricultural chemical factories.
Hexachlorocyclopentadiene	0.05	0.05	Kidney or stomach problems.	Discharge from chemical factories.
Lindane	0.0002	0.0002	Liver or kidney problems.	Runoff/leaching from insecticide used on cattle, lumber, gardens.
Methoxychlor	0.04	0.04	Reproductive difficulties.	Runoff/leaching from insecticide used on fruits, vegetables, alfalfa, livestock.
Oxamyl (Vydate)	0.2	0.2	Slight nervous system effects.	Runoff/leaching from insecticide used on apples, potatoes, and tomatoes.
Pentachlorophenol	zero	0.001	Liver or kidney problems; increased cancer risk.	Discharge from wood-preserving factories.
Picloram	0.5	0.5	Liver problems.	Herbicide runoff.
Polychlorinated biphenyls (PCBs)	zero	0.0005	Skin changes; thymus gland problems; immune deficiencies; reproductive or nervous system difficulties; increased risk of cancer.	Runoff from landfills; discharge of waste chemicals.
Simazine	0.004	0.004	Problems with blood.	Herbicide runoff.

Table continued next page.

Table 1-1 List of contaminants and their regulatory standard* (continued)

Contaminant	MCLG, *mg/L*†	MCL or TT, *mg/L*	Potential Health Effects From Ingestion of Water	Sources of Contaminants in Drinking Water
Styrene	0.1	0.1	Liver, kidney, or circulatory system problems.	Discharge from rubber and plastics factories; leaching from landfills.
Tetrachloroethylene	zero	0.005	Liver problems; increased risk of cancer.	Discharge from factories and dry cleaners.
Toluene	1	1	Nervous system, kidney, or liver problems.	Discharge from petroleum factories.
Toxaphene	zero	0.003	Kidney, liver, or thyroid problems; increased risk of cancer.	Runoff/leaching from insecticide used on cotton and cattle.
2,4,5-TP (Silvex)	0.05	0.05	Liver problems.	Residue of banned herbicide.
1,2,4-Trichlorobenzene	0.07	0.07	Changes in adrenal glands.	Discharge from textile finishing factories.
1,1,1-Trichloroethane	0.2	0.2	Liver, nervous system, or circulatory problems.	Discharge from metal degreasing sites and other factories.
1,1,2-Trichloroethane	0.003	0.005	Liver, kidney, or immune system problems.	Discharge from industrial chemical factories.
Trichloroethylene	zero	0.005	Liver problems; increased risk of cancer.	Discharge from metal degreasing sites and other factories.
Vinyl chloride	zero	0.002	Increased risk of cancer.	Leaching from PVC pipes; discharge from plastics factories.
Xylenes (total)	10	10	Nervous system damage.	Discharge from petroleum factories; discharge from chemical factories.
Radionuclides				
Alpha particles	none ——— zero	15 pCi/L	Increased risk of cancer.	Erosion of natural deposits of certain minerals that are radioactive and may emit a form of radiation known as alpha radiation.

Table continued next page.

Table 1-1 List of contaminants and their regulatory standard* (continued)

Contaminant	MCLG, *mg/L*†	MCL or TT, *mg/L*	Potential Health Effects From Ingestion of Water	Sources of Contaminants in Drinking Water
Beta particles and photon emitters	none ——— zero	4 millirems per year	Increased risk of cancer.	Decay of natural and synthetic deposits of certain minerals that are radioactive and may emit forms of radiation known as photons and beta radiation.
Radium-226 and radium-228 (combined)	none ——— zero	5 pCi/L	Increased risk of cancer.	Erosion of natural deposits.
Uranium	zero	30 µg/L as of 12/8/03	Increased risk of cancer; kidney toxicity.	Erosion of natural deposits.

*Definitions:

Maximum contaminant level (MCL)—The highest level of a contaminant that is allowed in drinking water. MCLs are set as close to MCLGs as feasible using the best available treatment technology and taking cost into consideration. MCLs are enforceable standards.

Maximum contaminant level goal (MCLG)—The level of a contaminant in drinking water below which there is no known or expected risk to health. MCLGs allow for a margin of safety and are nonenforceable public health goals.

Maximum residual disinfectant level (MRDL)—The highest level of a disinfectant allowed in drinking water. There is convincing evidence that addition of a disinfectant is necessary for control of microbial contaminants.

Maximum residual disinfectant level goal (MRDLG)—The level of a drinking water disinfectant below which there is no known or expected risk to health. MRDLGs do not reflect the benefits of the use of disinfectants to control microbial contaminants.

picocuries per liter—pCi/L.

N/A—Not applicable.

Treatment technique (TT)—A required process intended to reduce the level of a contaminant in drinking water.

†Units are in milligrams per liter (mg/L) unless otherwise noted. Milligrams per liter is equivalent to parts per million.

‡USEPA's Surface Water Treatment Rules (SWTRs) require systems using surface water or groundwater under the direct influence of surface water (GWUDI) to (1) disinfect their water and (2) filter their water or meet criteria for avoiding filtration so that the following contaminants are controlled at the following levels:

- *Cryptosporidium:* 99% removal by all filtered systems, plus additional treatment as required under the Long-Term 2 Enhanced Surface Water Treatment Rule of Jan. 5, 2006, including at least 99% inactivation by unfiltered surface water systems.
- *Giardia lamblia:* 99.9% removal/inactivation.
- Viruses: 99.99% removal/inactivation.

Table continued next page.

Table 1-1 List of contaminants and their regulatory standard* (continued)

- *Legionella*: No limit, but USEPA believes that if *Giardia* and viruses are removed/inactivated, *Legionella* will also be controlled.
- Turbidity: At no time can turbidity (cloudiness of water) go above 5 ntu; systems that filter must ensure that the turbidity go no higher than 1 ntu (0.5 ntu for conventional or direct filtration) in at least 95% of the daily samples in any month. As of Jan. 1, 2002, turbidity may never exceed 1 ntu and must not exceed 0.3 ntu for 95% of daily samples in any month.
- HPC: No more than 500 bacterial colonies per milliliter.
- Long-Term 1 Enhanced Surface Water Treatment Rule (effective date: Jan. 14, 2005): Surface water systems or GWUDI systems serving fewer than 10,000 people must comply with the applicable Long-Term 1 Enhanced Surface Water Treatment Rule provisions (e.g., turbidity standards, individual filter monitoring, *Cryptosporidium* removal requirements, updated watershed control requirements for unfiltered systems).
- Filter Backwash Recycling: The Filter Backwash Recycling Rule requires systems that recycle to return specific recycle flows through all processes of the system's existing conventional or direct filtration system or at an alternate location approved by the state.

§ No more than 5.0% samples total coliform-positive in a month. (For water systems that collect fewer than 40 routine samples per month, no more than 1 sample can be total coliform–positive per month.) Every sample that has total coliform must be analyzed for either fecal coliforms or *E. coli*; if two consecutive samples are total coliform–positive and one is also positive for *E. coli* fecal coliforms, the system has an acute MCL violation.

** Fecal coliform and *E. coli* are bacteria whose presence indicates that the water may be contaminated with human or animal wastes. Disease-causing microbes (pathogens) in these wastes can cause diarrhea, cramps, nausea, headaches, or other symptoms. These pathogens may pose a special health risk for infants, young children, and people with severely compromised immune systems.

†† Although there is no collective MCLG for this contaminant group, there are individual MCLGs for some of the individual contaminants:
Trihalomethanes: bromodichloromethane (zero); bromoform (zero); dibromochloromethane (0.06 mg/L); chloroform (0.07 mg/L).
Haloacetic acids: dichloroacetic acid (zero); trichloroacetic acid (0.02 mg/L); monochloroacetic acid (0.07 mg/L). Bromoacetic acid and dibromoacetic acid are regulated with this group but have no MCLGs.

‡‡ MCLGs were not established before the 1986 amendments to the SDWA. Therefore, there is no MCLG for this contaminant.

§§ Lead and copper are regulated by a treatment technique that requires systems to control the corrosiveness of their water. If more than 10% of tap water samples exceed the action level, water systems must take additional steps. The action level for copper is 1.3 mg/L; the action level for lead is 0.015 mg/L.

***Each water system must certify, in writing, to the state (using third-party or manufacturer's certification) that when acrylamide and epichlorohydrin are used in drinking water systems, the combination (or product) of dose and monomer level does not exceed the levels specified, as follows:
Acrylamide = 0.05% dosed at 1 mg/L (or equivalent)
Epichlorohydrin = 0.01% dosed at 20 mg/L (or equivalent).

Table 1-2 National Secondary Drinking Water Regulations

Contaminant	Secondary Standard
Aluminum	0.05 to 0.2 mg/L
Chloride	250 mg/L
Color	15 (color units)
Copper	1.0 mg/L
Corrosivity	Noncorrosive
Fluoride	2.0 mg/L
Foaming agents	0.5 mg/L
Iron	0.3 mg/L
Manganese	0.05 mg/L
Odor	3 threshold odor number
pH	6.5–8.5
Silver	0.10 mg/L
Sulfate	250 mg/L
Total dissolved solids	500 mg/L
Zinc	5 mg/L

NOTE: For more information, read *Secondary Drinking Water Regulations: Guidance for Nuisance Chemicals.*

Monitoring and Reporting Requirements

To ensure that drinking water meets federal and state requirements, all water systems are required to regularly sample and test the water supplied to consumers and to report results to the state. As detailed in Table 1-3, the regulations specify minimum sampling frequencies and sampling locations for various types of contaminants. In many cases, sampling regimes are a function of the type of water source being used, the type of treatment used, and the size of the water system.

In response to the increased number of regulated contaminants following the 1986 SDWA amendments, USEPA established a standardized monitoring framework in 1991 to reduce the variability within monitoring requirements for chemical and radiological contaminants across system sizes and types. As shown in Table 1-4, specific information must be included on every laboratory report. There are also specific requirements for the records that must be kept by water systems regarding operation and monitoring and for the length of time the records must be retained. These requirements are summarized in Table 1-5. Although state requirements for monitoring, reporting, and record retention must be as stringent as federal requirements, they often vary and may include specific procedures that must be used.

Table 1-3 Required sampling for regulated contaminants

Contaminant	Sampling Location	Frequency: Community and Nontransient, Noncommunity Systems	Frequency: Transient, Noncommunity Systems
Inorganics	Entry points to distribution system	Systems using surface water: every year Systems using groundwater only: every 3 years	Nitrate: yearly Nitrite: at state option
Organics: except regulated disinfection by-products	Entry points to distribution system	Systems using surface water: every 3 years Systems using groundwater only: state option	State option
Organics: regulated disinfection by-products*	25 percent at extremes of distribution system; 75 percent at locations presentative of population distribution	Systems serving populations of 10,000 or more: four samples per quarter per plant†	State option
Turbidity	At point(s) where water enters distribution system	Systems using surface water: daily Systems using groundwater only: state option	Systems using surface water or surface water and groundwater only: daily Systems using groundwater only: state option
Coliform bacteria	At consumer's faucet	Depends on number of people served by water system	Systems using surface water and/or groundwater: one per quarter (for each quarter water is served to public)
Radiochemicals: natural	At consumer's faucet	Systems using surface water: every 4 years Systems using groundwater only: state option	State option
Radiochemicals: synthetic	At consumer's faucet (at state option)	Systems using surface water serving populations greater than 100,000: every 4 years All other systems: state option	Systems using surface water and/or groundwater: state option

*Monitoring locations and frequencies will change in 2012 under the Stage 2 Disinfectants/Disinfection By-products Rule of Jan. 4, 2006.
†Systems using multiple wells drawing raw water from a single aquifer may, with state approval, be considered one treatment plant for determining the required number of samples.

Table 1-4 Laboratory report summary requirements

Type of Information	Summary Requirement
Sampling information	Date, place, and time of sampling
	Name of sample collector
	Identification of sample
	• Routine or check sample
	• Raw or treated water
Analysis information	Date of analysis
	Laboratory conducting analysis
	Name of person responsible
	Analytical method used
	Analysis results

Table 1-5 Record-keeping requirements

Type of Record	Time Period
Bacteriological and turbidity analyses	5 years
Chemical analyses	10 years
Actions taken to correct violations	3 years
Sanitary survey reports	10 years
Exemptions	5 years following expiration

If analysis of the water produced by a water system indicates that an MCL for a contaminant is exceeded, the water system must initiate a treatment regime to reduce the contaminant concentration to below the MCL or take appropriate steps to protect the public's health. The rules also identify BATs for treating contaminated water for specific contaminants, although their use is not required.

Notification Requirements

Public water systems that violate primary MCLs and treatment techniques as well as operating, monitoring, or reporting requirements must inform the public of the problem. Even though the problem may have already been corrected, an explanation must be provided in the news media describing the public health significance of the violation.

Some violations are more serious than others, and two tiers of public notification have been established (Tables 1-6 and 1-7). Tier 1 violations are more serious than Tier 2 violations and have more extensive notification requirements. USEPA provides mandatory language for use with each type of public notification to fully inform the public of the significance of the violation. Violations of state reporting requirements may not require public notification, depending on the infraction and state policy.

Table 1-6 Summary of notification requirements

Category of Violation	Mandatory Health Effects Information Required (all public water supplies)	Notice to New Billing Units (community water supplies only)
Tier 1		
Maximum contaminant level	Yes	Yes
Treatment technique	Yes	Yes
Variance of exemption schedule violation	Yes	Yes
Tier 2		
Monitoring*	No	No
Testing procedures	No	No
Variance of exemption issued	Yes	No

Source: Adapted from Public Notification Handbook for Public Officials *(USEPA 1989).*
*Continuous report required if posting is used; quarterly report required if hand delivery is used.

Water systems must also report monitoring data to the public annually (in the summer following a calendar monitoring year) in CCRs. As directed by USEPA in a 1998 regulation, this report must include other water system and health-risk information.

Public Water Systems

Only public water systems are subject to drinking water standards. Essentially, a public water system is any system that provides water through pipes or other conveyances for human consumption and has at least 15 service connections or regularly serves an average of at least 25 individuals daily for at least 60 days out of the year. As defined in the regulations, the system includes collection, treatment, storage, and distribution facilities under the control of the system operator and is used primarily in connection with the system. In addition, collection or pretreatment storage facilities that are not under the control of the water system operator, but that are used primarily in connection with the system, are considered part of the public water system. However, irrigation districts generally are not public water systems.

The SDWA excludes systems from consideration as public water systems if they:

- Consist only of distribution and storage facilities (and do not have any collection and treatment facilities)
- Obtain all of their water from, but are not owned or operated by, a public water system to which such regulations apply
- Do not sell water to any person
- Are not a carrier that conveys passengers in interstate commerce

Table 1-7 Types of notification and time frames for notification by community and noncommunity water supplies

Type of Public Water Supply	Type of Violation and Notification Method	Time Frame Within Which Notice Must Be Given (*X* indicates time frame for initial notice)					Frequency of Repeat Notices Until Violation Is Resolved
		72 hours	7 days	14 days	45 days	3 months	
Community	Tier 1: Acute violations						
	• TV and radio	X					No repeat
	• Newspaper*			X			No repeat
	• Mail or hand delivery†				X		Quarterly repeat
Community	Tier 1: Nonacute violations						
	• Newspaper*			X			No repeat
	• Mail or hand delivery				X		Quarterly repeat
Community	Tier 2: All violations						
	• Newspaper*					X	Quarterly repeat by mail or hand delivery
Noncommunity‡	Tier 1: Acute violations						
	• Option 1: Notice as for community water systems (above)						
	• Option: 2: Posting or hand delivery	X					Continuous or quarterly repeat
Noncommunity	Tier 1: Nonacute violations						
	• Option 1: Notice as for community water systems (above)						
	• Option 2: Posting or hand delivery			X			Continuous or quarterly repeat
Noncommunity	Tier 2: All violations						
	Option 1: Notice as for community water system (above)						
	Option 2: Posting or hand delivery					X	Continuous or quarterly repeat§

Source: Adapted from Public Notification Handbook for Public Officials *(USEPA 1989).*

*If no newspaper of general circulation is available, posting or hand delivery is required.

†May be waived.

‡Includes both transient, noncommunity public water systems and nontransient, noncommunity public water systems.

§Less frequent notice (but no less than annual) to be required.

To encourage water conservation, USEPA in 2003 issued guidance clarifying its interpretation of what constitutes a public water system. Under earlier agency interpretations, multiunit properties housing more than 25 people or comprising at least 15 units, such as apartment buildings, were subject to SDWA regulatory requirements if owners submeter water and bill for water use at individual units. Based on a finding that submetering encourages conservation because residents pay for their water use, the 2003 guidance excludes such properties as public water systems that merely submeter and bill for drinking water from a public water system already complying with the SDWA.

Classified by Type

Some drinking water regulations affect only community water systems (CWSs). A CWS is a public water system that serves at least 15 service connections used by year-round residents or regularly serves at least 25 year-round residents. Municipalities and rural water districts are typical examples of community water systems. If a public water system does not meet the definition of a CWS, it is a noncommunity water system (NCWS). There are two types of NCWSs:

1. Nontransient, noncommunity water systems (NTNCWSs)—These systems regularly serve at least 25 of the same persons over 6 months per year. Examples include schools, factories, and hospitals that have their own water supply.
2. Transient, noncommunity water system (TNCWSs)—These systems do not regularly serve at least 25 of the same persons over 6 months each year. Campgrounds, motels, and gas stations are examples of TNCWSs.

Because TNCWSs serve populations for relatively brief periods of time, they are least likely to be regulated under NPDWRs. However, regulations controlling contaminants that have the potential to cause acute illness during short exposures typically apply to TNCWSs.

As of 2004, water system inventory data show that CWSs account for 33 percent of all water systems and serve more than 273 million people. The data also indicate that NTNCWSs account for 12 percent of all systems and serve more than 6 million people while TNCWSs account for 55 percent of all systems and serve more than 23 million people.

Classified by Size

Within specific SDWA regulations, applicability may be contingent on the size of a system, or the stringency of compliance requirements may depend on a system's size. When this is the case, system size is usually based on the number of people served by the water system. For example, lead and copper treatment technique provisions depend on the size of the water system. Under these regulations, a small system is defined as one that serves no more than 3,300 people; a medium system serves between 3,301 and 50,000 people; and a large system serves more than 50,000 people. The vast majority of water systems serve fewer than 3,300 people. USEPA reports 84 percent of CWSs and nearly all NTNCWSs and TNCWSs are small.

System size cutoffs vary from regulation to regulation. Therefore, water system operators should always verify size-based applicability for specific regulations.

Classified by Source Type

A system's water source determines, in part, the type of contaminants that are likely to be found in the water. For example, bacteria are much more likely to occur in surface water than in groundwater. Thus, some microbial-control regulations apply only to systems using surface water and to systems using GWUDI, which is determined by states on a case-by-case basis.

Each source water type is defined as follows:

- Surface water is all water that is open to the atmosphere and subject to surface runoff.
- GWUDI generally includes any water beneath the surface of the ground with significant occurrence of insects or other microorganisms, algae, or large-diameter pathogens, such as *Giardia lamblia*, and any subsurface water that is subject to significant and relatively rapid shifts in water characteristics such as turbidity, temperature, conductivity, and pH that closely correlate to climatological or surface water conditions.
- Groundwater is not defined; by default, it includes water sources that are not surface water or GWUDI.

USEPA reports that 91 percent of all water systems rely on groundwater as their primary water source. However, the remaining 9 percent of water systems using surface water serve nearly twice as many people as systems using groundwater.

State Primacy

SDWA regulations apply to all 50 states, the District of Columbia, Native American lands, Puerto Rico, the Virgin Islands, American Samoa, Guam, the Commonwealth of the Northern Mariana Islands, and the Republic of Palau. The SDWA gives individual states the opportunity to set and enforce their own drinking water standards if the standards are at least as stringent as USEPA's national standards. Most states and territories directly oversee the water systems within their borders; this is referred to as primary enforcement authority, or primacy. States must submit their versions of federal regulations for USEPA approval within 2 years of promulgation.

Of the 57 states and territories, all but Wyoming and the District of Columbia had primacy as of February 2006. USEPA regional offices administer the public water supply supervision (PWSS) program within these two jurisdictions. Native American tribes also have the right to apply for and receive primacy.

Because some federal regulations give states flexibility to adapt the national requirements to state and local conditions and some states have adopted regulations that are more stringent than federal requirements, day-to-day implementation can be complicated.

Variances and Exemptions

Each drinking water regulation includes provisions for states to issue variances and exemptions. Variances allow utilities to deviate from an MCL or treatment technique under certain conditions; exemptions give utilities additional time to comply with a new regulation. Systems operating under either must provide drinking water that does not pose an unreasonable risk to public health, as determined by USEPA.

USEPA updated its variance and exemption regulations in 1998 to codify provisions in the 1996 SDWA amendments regarding affordability-based variances for systems serving fewer than 10,000 people. The revised regulations set forth requirements for general variances, which are available to public water systems of all sizes, and for small system variances. They also cover exemptions.

General Variances

The regulations authorize states to grant one or more general variances to a water system that cannot comply with an MCL because of characteristics of the water source(s). A general variance may only be granted to systems that have installed USEPA-designated BAT for treatment of the MCL being violated. Granting of a general variance must not result in an unreasonable risk to the public health, and the state must prescribe a schedule of compliance.

Small System Variances

States can grant small system variances to systems serving fewer than 3,300 people without USEPA approval but must get agency concurrence for variances to systems serving 3,301–10,000 people. However, such variances cannot be granted unless USEPA has identified an affordable variance technology for each particular regulation based on national affordability criteria and states determine that a system meets state-established affordability criteria. Under such criteria, variances can be granted only if states determine that a utility cannot afford to comply by using an alternative source or through restructuring or consolidation. To date, no such small system variances have been granted because USEPA has not identified any affordable small system variance technologies.

Exemptions

States may also exempt a water system from an MCL or treatment technique requirement if it finds that all three of the following conditions exist:

- The system is unable to comply with the requirement because of compelling factors, which may include economic factors.
- The exemptions will not result in an unreasonable risk to public health.
- The system was in operation as of Jan. 1, 1989, or, if it was not, no reasonable alternative source of drinking water is available to the new system.

CURRENT AND FORTHCOMING REGULATIONS

Primary drinking water regulations cover both microbial and chemical contaminants. Microbial contaminants are considered to pose immediate or acute public health risks, and chemical contaminants are considered to pose long-term or chronic health risks. Other regulations include the Unregulated Contaminant Monitoring Rule, CCR, and the Public Notice Rule.

Regulations to Control Microbial Contaminants

Major rules intended to control microbial risks include:

- Total Coliform Rule (TCR)
- Surface Water Treatment Rule (SWTR)
- Interim Enhanced SWTR (IESWTR)
- Long-Term 1 ESWTR (LT1ESWTR)
- Filter Backwash Recycling Rule (FBRR)
- Long-Term 2 ESWTR (LT2ESWTR)

Forthcoming rule intended to control microbial risks is the:

- Ground Water Rule (GWR)

Total Coliform Rule

The TCR, promulgated in June 1989 and effective as of January 1991, set both an MCLG and an MCL for total coliform bacteria in drinking water. The rule also details the type and frequency of testing that water systems must perform.

Coliforms are a broad class of bacteria that live in the digestive tracts of humans and many animals. The presence of coliform bacteria in tap water suggests that the treatment system is not working properly or that there is a problem in the distribution system pipes. Coliform contamination can cause diarrhea, cramps, nausea, and vomiting. Together these symptoms comprise a general category known as gastroenteritis. Gastroenteritis is not usually serious for a healthy person; however, it can lead to more serious problems for people with weakened immune systems, such as the very young, elderly, or immunocompromised.

In the rule, USEPA set the MCLG for total coliforms at zero. Because researchers have found very low levels of coliforms in specific waterborne disease outbreaks, any level of coliforms indicates some health risk. USEPA also set an MCL under which systems must not detect coliform bacteria in more than 5 percent of required monthly samples from the distribution system. If more than 5 percent of the samples contain coliforms, water system operators must report this violation to the state and the public.

Coliform bacteria in drinking water may indicate that the system's treatment system is not performing properly. To avoid or eliminate microbial contamination, systems may need to take a number of actions, including repairing the disinfection/filtration equipment, flushing or upgrading the distribution system, and enacting source water protection programs to prevent contamination.

If a sample tests positive for coliforms, the system must collect a set of repeat samples within 24 hours. When a routine or repeat sample tests positive for total coliforms, it must also be analyzed for fecal coliforms and *Escherichia coli* (*E. coli*), which are coliform bacteria directly associated with fecal contamination. A positive result to this last test signifies an acute MCL violation, which necessitates rapid state and public notification because it represents a direct health risk.

The number of coliform samples a system must take depends on the number of customers it serves. Systems serving fewer than 1,000 people may test once a month or less often; systems with 50,000 customers must test 60 times per month; and those with 2.5 million customers must test at least 420 times per month. These are minimum schedules, and many systems test more frequently.

Surface Water Treatment Rule

The SWTR, promulgated in June 1989 and effective as of January 1991, aims to prevent waterborne diseases caused by viruses, *Legionella*, and *Giardia lamblia* (a chlorine-resistant protozoan). These disease-causing microbes are present at varying concentrations in most surface waters. The rule requires water systems to treat water from surface water sources to reduce the occurrence of unsafe levels of these microbes.

Although it was strengthened under the IESWT, LT1ESWT, and LT2ESWT rules, the original SWTR still applies to the operation of every public water system that uses surface waters such as lakes and streams or groundwater determined to be GWUDI. The purpose of the regulation is to protect the public from waterborne diseases. The organisms that cause waterborne diseases most frequently diagnosed in the United States are *Giardia*, *Cryptosporidium*, *Legionella*, viruses, and some types of bacteria, including certain strains of *E. coli*. Although the SWTR does not directly address *Cryptosporidium*, another chlorine-resistant protozoan found in water sources, the rule's treatment requirements provide a measure of control for this organism, which is directly addressed under the IESWT and LT2ESWT rules.

Ingestion of *Cryptosporidium, Giardia,* and viruses can cause problems in the human digestive system, generally diarrhea, cramps, and nausea. *Legionella* bacteria in water are a health risk only if the bacteria are aerosolized (e.g., in an air-conditioning system or a shower) and inhaled. Inhalation can result in a type of pneumonia known as Legionnaires' disease.

Because no simple, inexpensive tests are available for detecting waterborne organisms such *Cryptosporidium, Giardia,* and *Legionella,* the SWTR established a treatment technique to control microbial contaminants. The rule also established MCLGs for *Legionella, Giardia,* and viruses at zero because any amount of exposure to these contaminants represents some health risk.

In recognition of the many treatment technologies used by and available to water systems to control microbial contaminants, the treatment technique sets treatment performance goals and allows utilities to select the appropriate compliance technologies that best meet local source water conditions and budgets. Given the resistance to chlorine disinfection of *Giardia* and viruses, the SWTR generally requires the use of both disinfection and filtration processes to inactivate and remove virtually all target microorganisms.

The SWTR presumes that properly designed and operated filtration and disinfection systems can achieve removal or inactivation of at least 99.9 percent of *Giardia* cysts and 99.99 percent of viruses. USEPA and water treatment researchers determined, however, that the removal values are best expressed as logarithm values, with 99.9 percent of *Giardia* cysts equating to 3 log and 99.99 percent of viruses to 4 log.

The rule also established strict criteria that surface water systems with pristine sources must meet to avoid filtration and achieve the rule's objectives through a combination of disinfection and watershed protection. The SWTR establishes separate requirements for filtered and unfiltered surface water systems.

Filtered Systems. Filtration technologies specified by the SWTR include conventional treatment, direct filtration, slow sand filtration, diatomaceous earth (DE) filtration, reverse osmosis, and alternate technologies, which include membrane technologies. For each type of filtration technology, USEPA has established specific log-removal credits based on research demonstrating removal efficiencies at existing treatment plants. As noted, the credits reflect the performance of well-designed and operated systems. To gain credits, utilities must demonstrate the adequacy of their filtration process by meeting USEPA-set turbidity limits in treated water.

According to the SWTR, systems using conventional or direct filtration must achieve finished water turbidity ≤0.5 ntu in at least 95 percent of their monthly measurements. Systems using slow sand or DE filtration generally must meet or beat 1 ntu in at least 95 percent of monthly samples. Regardless of the filtration technology, finished water under the SWTR must never exceed 5 ntu.

Filtered water systems must also provide disinfection to complement filtration and achieve the required log-removal values. Disinfection performance is measured under the SWTR by calculating what is known as a *C×T value,* where *C* is the disinfectant concentration and *T* is the time water is in contact with the disinfectant.

In general, effectiveness of a disinfectant in inactivating *Giardia* cysts and viruses depends on the:

- type of disinfectant used,
- residual concentration of the disinfectant,
- period of time the water is in contact with the disinfectant,

- water temperature, and
- pH of the water.

USEPA developed a series of tables of $C \times T$ values for various disinfectants that achieve certain log inactivation of the target organisms. Filtered systems must combine log-removal and log-inactivation values to achieve SWTR performance goals. Systems using filtration must calculate and meet the $C \times T$ values specified by their state primacy agency.

To ensure adequate microbial protection in the distribution system, the SWTR also requires water systems to provide continuous disinfection of the drinking water entering the distribution system and to maintain a detectable disinfectant residual level within the distribution system. The disinfectant residual of water entering the distribution system must be monitored continuously for systems serving more than 3,300, and the residual cannot be less than 0.2 mg/L for more than 4 hours during periods when water is being served to the public. Any time the residual falls below this level, the system must notify the state. The SWTR also requires disinfectant residuals to be measured at the same points in the distribution system that are used for coliform sampling.

Unfiltered Systems. To qualify for a filtration waiver under the SWTR, surface water systems must meet stringent criteria for source water quality and site-specific conditions, especially regarding the reliability and performance of disinfection processes and the effectiveness of watershed protection programs. Such systems generally must conduct extensive source water monitoring for coliform and turbidity levels and unfailingly meet the required $C \times T$ values to achieve the performance goals through log inactivation alone. All surface water systems without filtration must compute the $C \times T$ value for their treatment process daily, and this value must always be above the minimum value specified by USEPA.

Interim and Long-Term 1 Enhanced Surface Water Treatment Rules

Waterborne disease outbreaks linked to *Cryptosporidium*, particularly the 1993 outbreak in Milwaukee, Wis., raised concerns about the adequacy of the SWTR to protect the public against waterborne pathogens. The Milwaukee outbreak led to creation of a committee of stakeholders to advise USEPA in establishing tougher surface water control regulations. These regulations were meant to balance the need to control pathogens with the need to minimize disinfection by-products, which have been determined to pose their own public health risks. The advisory committee concluded that an enhanced SWTR should be developed; this led to promulgation of the IESWTR in December 1998. The new rule added a layer of requirements for systems that are subject to the SWTR and that serve more than 10,000 people. It became effective in January 2002.

The IESWTR includes specific treatment requirements for *Cryptosporidium* but does not relieve systems from continuing to meet SWTR requirements. Specifically, the IESWTR includes:

- an MCLG of zero for *Cryptosporidium*;
- a requirement to remove 99 percent (2 log) of *Cryptosporidium* for filtered systems;
- strengthened performance standards for combined filter effluent turbidity;
- requirements for monitoring of individual filter turbidity;
- provisions requiring utilities to profile and benchmark their disinfection performance;
- provisions subjecting GWUDI systems to the new rules dealing with *Cryptosporidium*;
- inclusion of *Cryptosporidium* in the watershed control requirements for unfiltered public water systems;

- requirements for covering new finished water reservoirs; and
- requirements for states to conduct periodic sanitary surveys for all surface water systems regardless of size.

The rule's tightened turbidity performance criteria and individual filter monitoring requirements are intended to optimize treatment reliability and enhance physical removal efficiencies to minimize *Cryptosporidium* levels in finished water. Turbidity requirements for combined filter effluent will remain at the current level, which is at least every 4 hours, according to the SWTR, while continuous monitoring is required for individual filters. For conventional and direct filtration systems, the filtered water turbidity must be ≤0.3 ntu in at least 95 percent of monthly measurements and must never exceed 1 ntu.

The IESWTR does not contain new turbidity provisions for slow sand or DE systems (they remain as they are under the SWTR). However, the rule allows systems to demonstrate to states that alternative filtration technologies, in combination with disinfection treatment, consistently achieve 99.9 percent removal and/or inactivation of *Giardia*, 99.99 percent removal and/or inactivation of viruses, and 99 percent removal of *Cryptosporidium*. For systems that so demonstrate, their filtered water turbidity must be less than or equal to a value determined by the state that indicates 2-log *Cryptosporidium* removal, 3-log *Giardia* removal, and 4-log virus removal in at least 95 percent of the measurements taken each month. Also, the turbidity level for such systems must not exceed a maximum value determined by the state. In addition, the rule includes disinfection profiling and benchmarking provisions to ensure continued levels of microbial protection while facilities take the steps necessary to comply with new DBP standards set forth in the companion Stage 1 DBP Rule.

In January 2002 USEPA promulgated the LT1ESWTR. This rule extends IESWTR-like requirements to reduce *Cryptosporidium* and turbidity to systems serving 10,000 or fewer people. No variances from the filtration and disinfection requirements of the SWTR are allowed. On a case-by-case basis, exemptions may be granted by states for all requirements except disinfection residual requirements at the point of entry to the distribution system.

Filter Backwash Recycling Rule

The FBRR is intended to reduce the possibility for recycle practices within the water treatment process to adversely affect the performance of the drinking water treatment plant. Also, the rule is intended to help prevent microbes, such as *Cryptosporidium*, from passing through treatment systems and into finished drinking water. Customers may become ill if they drink contaminated water.

Spent filter backwash water, thickener supernatant, and liquids from dewatering processes can contain microbial contaminants, often in very high concentrations. Recycling these streams can reintroduce microbes and other contaminants into the treatment system. Additionally, high-volume recycle streams may upset treatment processes, allowing contaminants to pass through the system. To minimize these risks, the FBRR requires that recycle streams pass through all existing conventional or direct filtration processes within the system that USEPA has recognized as capable of achieving 2-log *Cryptosporidium* removal. The FBRR also allows recycle streams to be reintroduced at an alternate location if that location is state approved.

Long-Term 2 Enhanced Surface Water Treatment Rule

Building on the foundations laid by the SWTR, IESWTR, and LT1ESWTR, the LT2ESWTR aims to improve control of *Cryptosporidium* in water systems that are more vulnerable to *Cryptosporidium* occurrence. USEPA has determined that these systems require additional treatment. This vulnerable subset of surface water systems includes filtered systems with the highest *Cryptosporidium* levels in their source waters and unfiltered systems. As promulgated in January 2006, the LT2ESWTR requires all surface water and GWUDI systems to conduct an initial round of source water monitoring for *Cryptosporidium* to determine their treatment requirements.

Large systems (those serving at least 10,000 people) must begin such monitoring 6 months following promulgation; monitoring will last for 2 years. Smaller systems will start monitoring after large systems complete their monitoring. These systems will initially conduct less-expensive monitoring for *E. coli* for 1 year and then conduct another year of *Cryptosporidium* monitoring only if their *E. coli* results exceed specified triggering concentrations. Systems may grandfather equivalent, previously collected data in lieu of conducting new monitoring. Also, systems are not required to monitor if they provide the maximum level of treatment required under the rule.

Filtered systems will be classified in one of four "risk bins" based on their monitoring results. Most systems are expected to be in the lowest bin, which carries no additional treatment requirements. Filtered systems classified in higher risk bins will be required to provide 90 to 99.7 percent (1.0- to 2.5-log) additional reduction of *Cryptosporidium* levels using a range of treatment and management strategies. These are listed in a "toolbox" and assigned certain removal credit values. Systems may select from toolbox technologies to meet their additional treatment requirements. All unfiltered systems must provide at least 99 or 99.9 percent (2- or 3-log) inactivation of *Cryptosporidium*, depending on their monitoring results.

The LT2ESWTR also requires utilities to profile their disinfection performance. This requirement ensures that they maintain adequate microbial control as they take steps to reduce the formation of DBPs under the companion Stage 2 Disinfection By-products Rule. This companion rule, also promulgated in January 2006, establishes more stringent monitoring requirements for certain DBPs. In addition, water systems will be required to address risks in uncovered finished water storage facilities by covering them or otherwise managing them.

Ground Water Rule

An additional microbial regulation is due to be promulgated by October 2006. The GWR will establish a process for systems using undisinfected groundwater to determine whether disinfection is needed.

Presently, only surface water and GWUDI systems are required to disinfect their water supplies. The 1996 SDWA amendments, however, require USEPA to require disinfection of groundwater systems "as necessary" to protect the public health.

As proposed, the GWR will establish multiple barriers to protect against bacteria and viruses in drinking water taken from public groundwater sources. It will also establish a targeted strategy to identify groundwater systems at high risk for fecal contamination and specify when corrective action (including disinfection) is required to protect consumers who receive groundwater from public systems. This rule will apply to public groundwater systems that have at least 15 service connections or regularly serve at least 25 individuals daily for at least 60 days out of the year. The rule will also apply to any system that mixes surface water and groundwater if the groundwater is added directly to the distribution system and provided to consumers without treatment.

Regulations to Control Chemical Contaminants

The major rules intended to control chemical risks include:

- Arsenic Rule
- Lead and Copper Rule (LCR)

- Stage 1 Disinfectants/Disinfection By-products Rule (Stage 1 D/DBPR)
- Stage 2 Disinfectants/Disinfection By-products Rule (Stage 2 D/DBPR)
- Radionuclides Rule

Forthcoming rule intended to control chemical risks is the:

- Radon Rule

Arsenic Rule

In 2001 USEPA finalized a regulation to reduce the public health risks from arsenic in drinking water. It revised the current drinking water standard for arsenic from 0.050 mg/L, set in 1942, to 0.010 mg/L, which becomes effective in January 2006. The rule also sets an arsenic MCLG of zero and requires monitoring for new systems and new drinking water sources. It also clarifies procedures for determining compliance with MCLs for previously regulated inorganic contaminants, synthetic organic contaminants, and volatile organic contaminants.

The revised Arsenic Rule applies to all community and nontransient, noncommunity water systems. USEPA estimates that just over 4,000 systems will have to install treatment to comply with the revised arsenic MCL.

Under the rule, arsenic monitoring requirements are consistent with monitoring for other IOCs regulated under the standardized monitoring framework. The 2005–2007 compliance period is the first monitoring period under the revised MCL. Because the Arsenic Rule allows grandfathered data and waivers, systems should not have to deviate from their current monitoring scheme. To satisfy the monitoring requirements, groundwater systems are required to sample once every 3 years and must complete sampling by Dec. 31, 2007. Surface water systems are required to sample annually and must complete sampling by Dec. 31, 2006.

Lead and Copper Rule

The LCR, promulgated in June 1991 and effective as of December 1992, is substantially different from other chemical regulations. The other rules require water systems to treat water so that it is clean and safe to drink when it leaves their facilities. This rule regulates two contaminants that nearly always taint drinking water after it leaves the treatment plant.

Lead and copper are both naturally occurring metals. Both have been used to make household plumbing fixtures and pipes for many years, though Congress banned the use of lead solder, pipes, and fittings in 1986. The two contaminants enter drinking water when water reacts with the metals in the pipes. This is likely to happen when water sits in a pipe for more than a few hours.

Lead and copper have different health effects. Lead is particularly dangerous to fetuses and young children because it can slow their neurological and physical development. Anemia is one sign of a child's exposure to high lead levels. Lead may also affect the kidneys, brain, nervous system, and red blood cells.

Copper is a health concern for several reasons. The human body requires very low levels of copper. However, in the short term, consumption of drinking water containing copper well above the action level can cause nausea, vomiting, and diarrhea. It can also lead to serious health problems in people with Wilson's disease. Exposure to drinking water containing copper above the action level over many years could increase the risk of liver and kidney damage.

To prevent these health effects, USEPA set MCLGs and a treatment technique for lead and copper. USEPA requires all community and nontransient, noncommunity water systems to evaluate not only the pipes in their distribution systems but also the age and types of housing that they serve.

Based on this information, systems must collect water samples at points throughout the distribution system that are vulnerable to lead contamination, including regularly used residential bathroom and kitchen taps. LCR monitoring requires water systems to collect first-draw samples at cold water taps in at-risk homes and buildings, with the number of sites based on system size. Systems must monitor every 6 months unless they qualify for reduced monitoring.

If 10 percent of required sampling shows lead levels above a 0.015-mg/L action level, the utility must take action to control corrosion and carry out public education to inform consumers of steps needed to reduce their exposure to lead. If lead levels continue to be elevated after anticorrosion treatment is installed, the utility must replace lead service lines. A similar action level (1.3 mg/L) and response plan covers copper. An action level is different from an MCL: an MCL is a legal limit on a contaminant, while an action level is a trigger for additional prevention or removal steps.

In regard to treatment, utilities must, at a minimum, maintain optimal corrosion control. Corrosion control does not reduce the contaminant level but helps prevent water contamination. By increasing the water's pH or hardness, water systems can make their water less corrosive and therefore less likely to corrode pipes and absorb lead or copper. Consumers can further reduce the potential for elevated lead levels at the tap by ensuring that all plumbing and fixtures meet local plumbing codes.

Public education under the LCR includes informing customers of health effects, sources of the contaminants, and steps they can take to limit their exposure. The LCR specifies places, ways, and schedules to deliver such information.

Based on an extensive national review of LCR implementation by states and utilities, USEPA initiated a Drinking Water Lead Reduction Plan in early 2005. The intent of the Plan is to clarify and strengthen certain aspects of the rule to improve its implementation and effectiveness. The plan specifically revises monitoring, treatment, lead service line management, and customer awareness requirements. It also updates guidance regarding lead in tap water in schools and child care facilities. USEPA intends to propose these and other regulatory changes in early 2006.

Stage 1 Disinfectants/Disinfection By-products Rule

Promulgated with the IESWTR in 1998, the Stage 1 D/DBPR applies to both surface and groundwater systems and has far-reaching effects for US water utilities. Its objective is to limit public exposure to DBPs, which are formed when organic materials, naturally present in source waters, combine with a disinfectant. Common organics present in many surface water sources are humic acids. Disinfectants commonly used in drinking water treatment include chlorine, chloramines, ozone, and chlorine dioxide. The amount and type of DBPs formed depend on many factors, including the amount and type of organic precursors initially present, pH, time of exposure to disinfectant, temperature, and type of disinfectant.

Unlike the existing total trihalomethane (TTHM) regulation, which applies only to systems serving more than 10,000 people, the Stage 1 Rule applies to all systems, regardless of the size of population served. The rule lowered the allowable TTHM level from 0.10 mg/L to 0.080 mg/L and added new MCLs for five haloacetic acids (HAA5) at 0.060 mg/L, chlorite at 1.0 mg/L, and bromate at 0.010 mg/L. It also established maximum residual disinfectant levels for chlorine (4 mg/L as Cl_2), chloramines (4.0 mg/L as Cl_2), and chlorine dioxide (0.8 mg/L as ClO_2).

The rule also includes a treatment requirement for "enhanced coagulation" for surface water sources where conventional treatment is applied. The removal of total organic carbon (TOC) to reduce the formation of DBPs is achieved using enhanced coagulation or enhanced softening. The rule specifies the percentage of influent TOC that must be removed based on the raw water TOC and alkalinity levels.

Stage 2 Disinfectants/Disinfection By-products Rule

The Stage 2 D/DBPR, promulgated in January 2006, supports its Stage 1 predecessor by requiring water systems to meet existing DBP MCLs at each monitoring site in the distribution system instead of averaging results from several sites. The new sampling sites are to be determined using a risk-targeting approach to identify locations where customers are exposed to the highest levels of regulated DBPs. The goal is to reduce DBP exposure among all customers.

Under the rule, most water systems must implement an Initial Distribution System Evaluation (IDSE) Program to identify compliance monitoring sites with the highest DBP levels. Long-term compliance monitoring, which begins in April 2012 for the largest systems, will then be determined by meeting TTHM and HAA5 MCLs at each site on a locational running annual average basis. Under the Stage 1 Rule, compliance is determined on a single systemwide running annual average of all samples.

The Stage 2 D/DBPR also requires systems to determine if they are experiencing short-term peaks in DBP levels, referred to as operational evaluation levels. Systems experiencing such peaks must review their operational practices to determine actions to be taken to prevent future excursions.

The rule applies to all community and nontransient, noncommunity systems that add a primary or residual disinfectant other than UV light or deliver water that has been disinfected by a primary or residual disinfectant other than UV. The rule is the first to recognize the efficacy of UV for inactivating chlorine-resistant microbial contaminants, such as *Cryptosporidium* and *Giardia*. The rule is accompanied by guidance on selecting and operating appropriate UV treatment technology.

The Stage 2 D/DBPR also adopts a standardized national approach for regulating interconnected consecutive and wholesale systems. The approach requires all utilities in such a combined distribution system to adhere to the compliance schedule based on the population served by the largest utility in the interconnected system, starting with the IDSE, which each utility must conduct on its own. However, each utility in a combined system will determine the number of samples to collect and the frequency for both the IDSE and Stage 2 D/DBPR routine compliance monitoring based on its own service population.

Radionuclides Rule

USEPA promulgated revised drinking water standards for radionuclide contaminants (other than radon) in December 2000, updating regulations first set in 1976. The regulation retained existing standards for combined radium-226/228, (adjusted) gross alpha, beta particle, and photon radioactivity. It also set a new standard for uranium (30 µg/L) as required by the SDWA and an MCLG of zero for all radionuclides. In addition, the rule promulgated separate monitoring requirements for radium-228 to ensure compliance with the combined radium-226/228 standard.

Under the new rule, effective as of December 2003, community water systems must meet the final MCLs and requirements for monitoring and reporting. The rule requires all new monitoring to be conducted at each entry point to the distribution system under a schedule designed to be consistent with the Standardized Monitoring Framework (SMF). The new monitoring requirements are being phased in by states through the start of the next SMF period, which is Dec. 31, 2007.

The rule requires water systems to determine initial compliance under the new monitoring requirements using the average of four quarterly samples or, at state discretion, using appropriate grandfathered data. Compliance will be determined immediately based on the annual average of the quarterly samples for that fraction of systems required by the state to monitor in any given year or based on the results from the grandfathered data.

Under the 1976 rule, water systems with multiple entry points to the distribution system were not required to test at every entry point but rather at a "representative point" in the distribution system. The new rule requires monitoring at all entry points to ensure that all customers receive water that meets the MCLs.

Radon Rule

There is one major forthcoming rule to regulate chemical contaminants. The long-delayed Radon Rule was first proposed in 1999 and is now expected to be promulgated in late 2007 at the earliest.

As proposed, the rule reflects a framework set forth in the 1996 SDWA amendments that provides for a multimedia approach to address the public health risks from radon in drinking water and in indoor air from soil. The framework reflects the fact that radon released to indoor air from soil under homes and buildings is the main source of exposure. Radon released from tap water is a much smaller source of radon in indoor air, which makes it more cost-effective to reduce the risk from radon in indoor air.

The proposed rule would establish an MCL of 300 pCi/L for radon in drinking water. This MCL would have to be met by systems that do not implement a broad program to mitigate radon in air and water. Water systems that implement such a multimedia mitigation program would have to meet a radon MCL of 4,000 pCi/L.

The proposed rule would apply to all community water systems that use groundwater or mixed ground and surface water, such as systems serving homes, apartment buildings, and trailer parks.

Other Regulations

In addition to regulations covering specific chemical and microbiological contaminants, USEPA has promulgated several programmatic regulations that address wider issues such as public notification and monitoring of unregulated contaminants. These are summarized in relevant fact sheets and quick reference guides in chapter 2 as follows:

- Unregulated Contaminant Monitoring Rule (p. 113)
- Variances and Exemptions Rule (p. 118)
- Public Notification Rule (p. 123)
- Consumer Confidence Report Rule (p. 127)
- Standardized Monitoring Framework for Chemical Contaminants (p. 130)

SELECTED SUPPLEMENTAL RESOURCES

USEPA's Office of Ground Water and Drinking Water web site (www.epa.gov/safewater) has the latest updates on drinking water regulatory issues, including extensive implementation guidance. Also, most state drinking water primacy agencies have web sites with the latest state regulatory information. All USEPA regulatory actions are published in the *Federal Register*, which is available on the Internet (www.gpoaccess.gov/fr/browse.html) and in most libraries.

AWWA publishes helpful regulatory information in its books, manuals, periodicals, and videos and provides relevant training and technical programs, which can be found online at www.awwa.org.

SDWA Rules and Regulations

This chapter includes easy-to-use summaries of SDWA rules and regulations, which are based on the US Environmental Protection Agency's series of Quick Reference Guides. To simplify use of this chapter, acronyms and abbreviations are not defined within the text. Refer to the list of acronyms and abbreviations at the end of the book. The chapter begins with a table that lists the key points for current rules. It is followed by a table that lists key points for proposed/upcoming rules. These tables are followed by quick reference guides for the following:

Regulations to Control Microbial Contaminants
- Total Coliform Rule
- Comprehensive Surface Water Treatment Rules (Except LT2ESWTR)—Systems Using Slow Sand, Diatomaceous Earth, or Alternative Filtration
- Comprehensive Surface Water Treatment Rules (Except LT2ESWTR)—Systems Using Conventional or Direct Filtration
- Comprehensive Surface Water Treatment Rules (Except LT2ESWTR)—Unfiltered Systems
- Interim Enhanced Surface Water Treatment Rule
- Long-Term 1 Enhanced Surface Water Treatment Rule
- Long-Term 2 Enhanced Surface Water Treatment Rule
- Filter Backwash Recycling Rule
- Proposed Ground Water Rule

Regulations to Control Chemical Contaminants
- Arsenic and Clarifications to Compliance and New Source Monitoring Rule
- Lead and Copper Rule
- Stage 1 Disinfectants/Disinfection By-products Rule
- Stage 2 Disinfectants/Disinfection By-products Rule
- Radionuclides Rule
- Proposed Radon in Drinking Water Rule

Other Regulations
- Unregulated Contaminant Monitoring Rule
- Variances and Exemptions
- Public Notification Rule
- Consumer Confidence Report Rule
- Standardized Monitoring Framework

Key Points: Regulations to Control Microbial Contaminants

Rule	Systems Affected	Overview
Total Coliform Rule *Published: June 29, 1989*	All PWSs.	Sets monitoring requirements for coliforms, which are indicators of the potential for sewage or fecal contamination in the water.
Surface Water Treatment Rule *Published: June 29, 1989*	All PWSs that use surface water or GWUDI. These are defined as Subpart H systems.	Establishes criteria under which filtration is required. Systems must either provide filtration and disinfection or comply with the requirements to avoid filtration.
Long-Term 1 Enhanced Surface Water Treatment Rule *Published: Jan. 14, 2002* **Interim Enhanced Surface Water Treatment Rule** *Published: Dec. 16, 1998*	The Interim ESWTR applies to all PWSs that use surface water or GWUDI and serve at least 10,000 people. In addition, it requires states to conduct sanitary surveys for all surface water and GWUDI systems, including those that serve fewer than 10,000 people. The Long-Term 2 ESWTR applies to all PWSs that use surface water or GWUDI and serve fewer than 10,000 people.	Phased in by system size, both rules build on SWTR requirements by strengthening control of microbial contaminants, especially *Cryptosporidium*. They aim to optimize treatment reliability by establishing tighter turbidity performance criteria for filtered systems, requiring filtered systems to achieve 2-log removal of *Cryptosporidium*, and including *Cryptosporidium* control in watershed control requirements for unfiltered systems. In addition, they require systems to conduct disinfection profiling and benchmarking to prevent increases in microbial risk while controlling for DBPs. The Interim ESWTR rule also banned new construction of uncovered finished water reservoirs.

Table continued next page.

Key Points: Regulations to Control Microbial Contaminants (continued)		
Monitoring	**Treatment**	**Management Practices**
For small systems, the number of monthly samples is based on service population. Repeat samples are required within 24 hours if a positive total coliform sample is found. Positive samples must be analyzed for *E. coli* or fecal coliform. At least five samples taken from sites in the distribution system must be collected the month after a positive sample.	This rule does not directly require implementation of treatment. However, if monitoring indicates the presence of coliform bacteria, treatment may have to be added, modified, or adjusted to correct the problem.	This rule does not directly affect a system's management practices. However, management practices may need to be adjusted to meet the monitoring and reporting requirements and/or to address any problems that are uncovered during monitoring.
For systems that *do not* provide filtration, the following samples are required: Source water: *Fecal or total coliform density:* 1–3 times per week, depending on the number of people served. *Turbidity:* Every 4 hours. Finished water: *Total inactivation ratios:* Daily. *Residual disinfectant concentration:* Continuously. **For systems that *do* provide filtration,** **the following samples are required:** *Turbidity:* Every 4 hours. *Residual disinfectant concentration:* Continuously.	Systems may avoid filtration if they have low coliform and turbidity in their source water and meet other site-specific criteria. Systems that do not meet these criteria must install one of the following filtration treatments: conventional filtration treatment or direct filtration; slow sand filtration; diatomaceous earth filtration; or another filtration technology if the state determines that, in combination with disinfection, the proper amount of *Giardia* and virus removal and/or inactivation is achieved.	Unfiltered systems are required to meet source water quality criteria and maintain a watershed control program. They are also subject to an annual inspection and watershed control program evaluation.
Continuous *turbidity monitoring* will be required for each individual filter (conventional and direct filtration only), and values will need to be recorded every 15 minutes. This is in addition to monitoring a combined flow from all filters established under the SWTR. This will prevent the situation in which a properly working filter masks the poor performance of another filter, thereby allowing contaminants to enter the water.	Performance standards of conventional and direct filtration plants also become more strict under this rule. Combined filter effluent must be ≤0.3 ntu for 95 percent of the monthly readings and may at no time exceed 1 ntu.	Management at those systems required to comply must establish a *disinfection profile and benchmark.* If a system is considering making a significant change in its disinfection practices (e.g., to comply with new Disinfectants/Disinfection By-products rules), it must get approval from the state. The state will use the benchmark as a guideline in deciding the level of disinfection that the system will need to achieve with its new disinfection practices.

Table continued next page.

Key Points: Regulations to Control Microbial Contaminants (continued)		
Rule	**Systems Affected**	**Overview**
Long-Term 2 Enhanced Surface Water Treatment Rule* *Published: Jan. 5, 2006*	All PWSs that use surface water or GWUDI (Subpart H systems).	The LT2ESWTR was promulgated concurrently with the Stage 2 D/DBPR to ensure that microbial protection is not compromised by efforts to reduce exposure to DBPs. It is also designed to require higher levels of treatment for source waters of lower quality.
Filter Backwash Recycling Rule *Published: June 8, 2001*	PWSs that use surface water or GWUDI (Subpart H systems), use conventional or direct filtration, and recycle spent filter backwash water, thickener supernatant, or liquids from dewatering processes.	The FBRR requires systems to return regulated streams to a point in the treatment plant where it goes through all of the steps of a conventional or direct filtration system. This is designed to ensure that inadequately treated water is not passed on to the distribution system and then to customers.
Ground Water Rule *Date Proposed: May 10, 2000* Expected to be published late 2006	PWSs that use groundwater.	The proposed Ground Water Rule aims to protect people served by groundwater systems from disease-causing viruses and bacteria. It will also seek to identify defects in water systems that could lead to contamination.

*This rule was promulgated simultaneously with the Stage 2 D/DBPR in order to protect public health and optimize technology choice decisions.

Table continued next page.

Key Points: Regulations to Control Microbial Contaminant (continued)

Monitoring	Treatment	Management Practices
NOTE: Monitoring takes place at the source prior to treatment. For small systems (serving fewer than 10,000 persons): Under the final rule, *Cryptosporidium* monitoring would be required if *E. coli* annual mean concentrations reach certain levels.	Depending on the initial monitoring results, systems that filter would be put into groups, or "bins." Under the final rule, each bin (except the bin for the lowest levels) requires a system to install additional treatment technology and sets a monitoring schedule, both based on contamination levels in the source water. Some new treatment options could possibly involve watershed control, reducing influent *Cryptosporidium* concentrations, improving system performance, and additional treatment barriers such as pretreatment.	This rule does not directly address management practices. However, should the installation of new treatment technology or the adoption of new treatment options be required, some management practices may be affected.
This rule has no monitoring requirements, but the FBRR requires that a system meet the following deadlines: **By Dec. 8, 2003:** Submit a plant schematic and recycle flow/plant flow information to the state. **By June 8, 2004:** Retain additional information on recycle practices on file. By this date systems must also be recycling regulated streams to correct locations or have an approved alternate recycle return location. **By June 8, 2006:** Any capital improvements that were needed for the return recycle location must be completed.	The FBRR requires a system to return all regulated recycled water (spent filter backwash water, thickener supernatant, and liquids from dewatering processes) to a point in the treatment plant where it will pass through all steps of treatment or treatment processes before entering the distribution system. NOTE: Systems can request approval from the state to use alternate locations.	This rule does not directly affect a system's management practices. However, management practices may need to be adjusted along with any change in treatment that is required.
Under the proposed rule, systems that do not achieve a high enough level of virus removal and/or inactivation must, after a positive total coliform result, take a source water sample and conduct further tests (e.g., for *E. coli,* enterococci, or coliphage). Under the proposed rule, states would conduct hydrogeologic sensitivity assessments, and systems identified as being sensitive will have further source water monitoring requirements.	This proposed rule does not directly affect treatment. However, systems that detect fecal contamination would be required to take corrective action that may include disinfection.	This proposed rule does not directly address management practices. However, states would evaluate system management as part of sanitary surveys and may require changes.

Key Points: Regulations to Control Chemical Contaminants		
Rule	**Systems Affected**	**Overview**
Phase I Rule *Published: July 8, 1987* **Phase II Rule** *Published: Jan. 30, 1991* **Phase IIB Rule** *Published: July 1, 1991* **Phase V Rule** *Published: July 17, 1992*	In general, requirements apply only to CWSs and NTNCWSs. Nitrate and nitrite requirements apply to all PWSs, including transient systems.	Establishes monitoring requirements and MCLs or treatment techniques for 66 chemicals (IOCs, VOCs, and SOCs).*
Arsenic and Clarifications to Compliance and New Source Contaminants Monitoring Rule *Revised Rule Published: Jan. 22, 2001*	CWSs and NTNCWSs. (NTNCWSs were not regulated under the previous rule.)	The Arsenic Rule sets an MCL as well as monitoring requirements for arsenic, a contaminant shown to cause cancer and other health effects. The revised rule reduces the MCL from the current 0.05 mg/L to 0.010 mg/L.

Table continued next page.

Key Points: Regulations to Control Chemical Contaminants (continued)

Monitoring	Treatment	Management Practices
The **Standardized Monitoring Framework.** Promulgated under the Phase II Rule, standardizes monitoring requirements and synchronizes monitoring schedules for IOCs, VOCs, and SOCs. Monitoring requirements for asbestos, fluoride, nitrate, and nitrite are different from the monitoring requirements for other IOCs because these chemicals have unusual characteristics. The SMF established a 9-year "compliance cycle" composed of three 3-year "compliance periods." Newly regulated contaminants will be subject to the SMF. During an initial monitoring period, systems sample for four consecutive quarters for each contaminant at each entry point to the distribution system. Depending on the results, systems may be able to reduce their monitoring frequency to annually or once every 3, 6, or 9 years. The SMF allows states to waive monitoring requirements for all contaminants except nitrate and nitrite.	These rules do not directly affect a system's treatment processes. However, if monitoring indicates chemical contamination, treatment may have to be added, modified, or adjusted to correct the problem.	This rule does not directly affect a system's management practices. However, management practices may need to be improved to meet the monitoring and reporting requirements and/or to address any problems that are uncovered during monitoring.
The final Arsenic Rule makes the monitoring requirements for arsenic consistent with those for other IOCs regulated under the SMF. Your state will set up a monitoring schedule that will allow you to monitor for all IOCs, including arsenic, at the same time.	This rule lists BATs and SSCTs for the removal of arsenic. The BATs and SSCTs that are most likely to be used by small systems include activated alumina, activated alumina and reverse osmosis POU devices, and modified lime softening.	This rule does not directly address management practices. However, systems that are required to install treatment for the first time will need to focus on developing appropriate technical, managerial, and financial capacity. Systems opting for a POE or POU compliance strategy will need to establish and maintain excellent customer relations.

Table continued next page.

Key Points: Regulations to Control Chemical Contaminants (continued)

Rule	Systems Affected	Overview
Lead and Copper Rule *Published: June 7, 1991*	All CWSs and NTNCWSs.	Establishes a treatment technique that includes requirements for corrosion control treatment, source water treatment, lead service line replacement, and public education. These requirements may be triggered by lead and copper action levels measured in samples collected at consumers' taps.
Stage 1 Disinfectants/ Disinfection By-products Rule *Published: Dec. 16, 1998*	CWSs and NTNCWSs that add a chemical disinfectant to the water in any part of the drinking water treatment process. Certain requirements apply to TNCWSs that use chlorine dioxide.	The Stage 1 D/DBPR will reduce the levels of disinfectants and DBPs in drinking water supplies, including by-products that were not previously covered by drinking water rules. DBPs result from chemical reactions between chemical disinfectants and organic and inorganic compounds in source waters. The rule sets MCLs for HAAs, chlorite (a major chlorine dioxide by-product), bromate (a major ozone by-product), and TTHMs. It also sets maximum residual disinfectant levels and maximum residual disinfectant level goals for chlorine, chloramines, and chlorine dioxide.
Stage 2 Disinfectants/ Disinfection By-products Rule† *Published: Jan. 4, 2006*	CWSs and NTNCWSs that add a disinfectant other than UV light or deliver water that has been disinfected.	The Stage 2 D/DBPR builds on the public health protection provided by the Stage 1 D/DBPR. Along with the proposed LT2ESWTR, it aims to reduce the risks associated with DBPs without increasing the risk of microbial contamination.

Table continued next page.

Key Points: Regulations to Control Chemical Contaminants (continued)		
Monitoring	**Treatment**	**Management Practices**
Samples must be taken from consumers' taps. The number of samples required during each 6-month period depends on system size. If monitoring results show that the lead or copper action level is exceeded, the system must implement corrosion control treatment. If the system is below the action level for two consecutive periods, it will be put on a reduced monitoring schedule.	Corrosion control treatment is required unless a system is below the action level for two consecutive 6-month periods. Source water monitoring and treatment are also required if a system exceeds an action level due to occurrence in the source water. If a system has lead service lines, replacement is required if the system still cannot meet the action level even after installing corrosion control or source water treatment.	This rule does not directly affect a system's management practices. However, management practices may be affected by the rule's public education provisions and practices associated with proper monitoring.
Depending on the type of disinfection used—chlorine, chloramines, chlorine dioxide, or ozone—systems may be required to monitor for different disinfectants and DBPs. Reduced monitoring is possible if a system meets certain requirements. For systems that use surface water or GWUDI (Subpart H systems) and use conventional treatment, monthly samples are required for TOC and alkalinity.	Subpart H systems that use conventional filtration must remove specified percentages of TOC using either enhanced coagulation or enhanced softening. The removal requirement depends on the TOC concentration in and alkalinity of the source water.	This rule does not directly affect a system's management practices. However, management practices may be affected by the need to balance disinfection needs with by-product formation.
Under the final rule, an *IDSE* will determine where the new monitoring sites will be located. The monitoring schedule would be based on both source water type and system size.	This rule may directly cause changes in treatment. In order to reduce DBP concentrations in the distribution system, systems may need to make operational changes or distribution system modifications, use alternative disinfection strategies, enhance DBP precursor removal, and/or remove DBPs.	This rule does not directly address management practices. However, should the installation of a new treatment technology or distribution system modifications be required, some management practices may be affected.

Table continued next page.

Key Points: Regulations to Control Chemical Contaminants (continued)

Rule	Systems Affected	Overview
Radionuclides Rule *Revised Rule Published: Dec. 7, 2000*	CWSs.	The Radionuclides Rule sets MCLs as well as monitoring, reporting, and public notification requirements for radionuclides, which are contaminants that emit radiation. The new rule maintains the current MCLs (from the original 1976 rule) for radium-226, radium-228, and gross alpha. Changes include establishing a new MCL for uranium, requiring systems to monitor separately for radium-228, and requiring systems to monitor for the regulated radionuclides at each entry point to the distribution system.
Radon Rule *Date Proposed: Nov. 2, 1999* Expected to be published late 2007	CWSs that use groundwater, mixed ground and surface water, GWUDI, or that intermittently use groundwater as a supplemental source. It will not apply to systems that rely on surface water exclusively.	The proposed Radon Rule aims to reduce people's exposure to radon in drinking water and in indoor air. Under the proposed rule, states would have the option to develop a multimedia mitigation program to address radon in both indoor air as well as drinking water.
*Fluoride is regulated with the other IOCs. However, in addition to a primary MCL of 4 mg/L, it has a secondary MCL of 2 mg/L. If your system has fluoride levels between 2 and 4 mg/L, you are required to provide public education about possible cosmetic dental discoloration.		
†This rule was promulgated simultaneously with the LT2ESWTR in order to protect public health and optimize technology choice decisions.		

Table continued next page.

Key Points: Regulations to Control Chemical Contaminants (continued)

Monitoring	Treatment	Management Practices
Monitoring for gross alpha, radium-226, radium-228, and uranium fit into the SMF. Monitoring will be required at each entry point to the distribution system. Monitoring for beta particle and photon emitters is not required for most CWSs. If a system is designated by the state as "vulnerable" or "contaminated," it will have to monitor for beta particle and photon radioactivity.	This rule lists BATs for the removal of radionuclides, should a capital investment be required. The BATs are ion exchange, reverse osmosis, lime softening, and enhanced coagulation/filtration. The SSCTs listed in the Radionuclides Rule are green sand filtration, coprecipitation with barium sulfate, electrodialysis/electrodialysis reversal, preformed hydrous manganese oxide filtration, activated alumina, and POE and POU devices, including POU ion exchange and POU reverse osmosis.	This rule does not directly address management practices. However, the rule involves new monitoring requirements, which may require improved management. In addition, should the installation of a treatment process be required, appropriate management practices may need to be implemented.
Under the proposed rule, the results of an initial monitoring period would determine the frequency of further monitoring that will be required. Sampling frequencies may be decreased if a system meets certain requirements or increased if sampling results exceed radon trigger levels.	Under the proposed rule, treatment technologies that are considered for radon treatment include high-performance aeration (pretreatment and posttreatment may also be necessary to avoid bacteriological growth and distribution system corrosion), granular activated carbon, and POE granular activated carbon (POU devices are not allowed for radon removal). Special consideration for spent media or cartridge disposal may be required if radon accumulates to high levels in the media.	This proposed rule does not directly address management practices. However, should monitoring be required, some management practices may be affected.

Key Points: Other Regulations

Rule	Systems Affected	Overview
Public Notification Rule *Published: May 4, 2000*	All PWSs.	The PN Rule ensures that all people who drink a system's water are informed about any violations that have occurred and their possible health consequences. The rule groups the public notification requirements into three tiers, depending on the seriousness of the violation or situation: *Tier 1* violations and situations have serious health effects with even a short-term exposure. Systems must issue notice within 24 hours. *Tier 2* violations and situations have the potential for serious effects on human health, though not as immediate as Tier 1. Notice is required within 30 days. *Tier 3* violations and situations do not present an immediate or serious risk. Notice is required within the year. The PN Rule also specifies how these notices are to be delivered.
Consumer Confidence Report Rule *Published: Aug. 19, 1998*	All CWSs.	The CCR is required to keep customers informed about the quality of their drinking water. A CCR is a report of water quality over the preceding year and includes health effects information. It includes information on source water, contaminants found in the water, and violations.

Key Points: Other Regulations (continued)

Monitoring	Treatment	Management Practices
This rule does not directly involve monitoring. However, the rule requires that certain monitoring results from other rules be reported to the public.	This rule does not directly affect treatment. However, treatment problems affect water quality and may cause violations that must be reported to the public.	PWSs must notify everyone they serve any time they fail to comply with the NPDWRs and in certain other circumstances.
This rule does not directly involve monitoring. However, the rule requires that certain monitoring results from other rules be reported in the CCR.	This rule does not directly affect treatment. However, treatment problems affect water quality, which must be reported in the CCR.	CWSs are required to make a CCR available annually to all customers.

Summaries of Regulations to Control Microbial Contaminants

TOTAL COLIFORM RULE

Overview	
Title	**Total Coliform Rule (TCR): 54 FR 27544–27568, June 29, 1989, Vol. 54, No. 124***
Purpose	Improve public health protection by reducing fecal pathogens to minimal levels through control of total coliform bacteria, including fecal coliforms and *E. coli.*
General Description	Establishes a MCL based on the presence or absence of total coliforms, modifies monitoring requirements including testing for fecal coliforms or *E. coli,* requires use of a sample siting plan, and also requires sanitary surveys for systems collecting fewer than *five* samples per month.
Utilities Covered	The TCR applies to all public water systems.
*The June 1989 rule was revised as follows: Corrections and Technical Amendments, 6/19/90 and Partial Stay of Certain Provisions (Variance Criteria) 56 FR 1556–1557, Vol. 56, No. 10. NOTE: The TCR is currently undergoing the 6-year review process and is subject to change.	

Public Health Benefits	
Implementation of the TCR has resulted in...	Reduction in risk of illness from disease-causing organisms associated with sewage or animal wastes. Disease symptoms may include diarrhea, cramps, nausea, and, possibly, jaundice, and associated headaches and fatigue.

What Are the Major Provisions?
ROUTINE Sampling Requirements
• Total coliform samples must be collected at sites that are representative of water quality throughout the distribution system according to a written sample siting plan subject to state review and revision. • Samples must be collected at regular time intervals throughout the month except groundwater systems serving 4,900 persons or fewer may collect them on the same day. • Monthly sampling requirements are based on population served (see table on next page for the minimum sampling frequency). • A reduced monitoring frequency may be available for systems serving 1,000 persons or fewer and using only groundwater if a sanitary survey within the past 5 years shows the system is free of sanitary defects (the frequency may be no less than 1 sample/quarter for community and 1 sample/year for noncommunity systems). • Each total coliform-positive routine sample must be tested for the presence of fecal coliforms or *E. coli.* • If any routine sample is total coliform-positive, repeat samples are required.

Table continued next page.

What Are the Major Provisions? (continued)
REPEAT Sampling Requirements
• Within 24 hours of learning of a total coliform-positive ROUTINE sample result, at least three REPEAT samples must be collected and analyzed for total coliforms: — One REPEAT sample must be collected from the same tap as the original sample. — One REPEAT sample must be collected within five service connections upstream. — One REPEAT sample must be collected within five service connections downstream. — Systems that collect one ROUTINE sample per month or fewer must collect a fourth REPEAT sample. • If any REPEAT sample is total coliform-positive: — The system must analyze that total coliform-positive culture for fecal coliforms or *E.coli.* — The system must collect another set of REPEAT samples, as before, unless the MCL has been violated and the system has notified the state.
Additional ROUTINE Sample Requirements
• A positive ROUTINE or REPEAT total coliform result requires a minimum of five ROUTINE samples be collected the following month the system provides water to the public unless waived by the state.

Public Water System ROUTINE Monitoring Frequencies

Population	Minimum Samples/ Month	Population	Minimum Samples/ Month	Population	Minimum Samples/Month
25–1,000*	1	21,501–25,000	25	450,001–600,000	210
1,001–2,500	2	25,001–33,000	30	600,001–780,000	240
2,501–3,300	3	33,001–41,000	40	780,001–970,000	270
3,301–4,100	4	41,001–50,000	50	970,001–1,230,000	300
4,101–4,900	5	50,001–59,000	60	1,230,001–1,520,000	330
4,901–5,800	6	59,001–70,000	70	1,520,001–1,850,000	360
5,801–6,700	7	70,001–83,000	80	1,850,001–2,270,000	390
6,701–7,600	8	83,001–96,000	90	2,270,001–3,020,000	420
7,601–8,500	9	96,001–130,000	100	3,020,001–3,960,000	450
8,501–12,900	10	130,001–220,000	120	3,960,001	480
12,901–17,200	15	220,001–320,000	150		
17,201–21,500	20	320,001–450,000	180		

*Includes public water systems that have at least 15 service connections but serve fewer than 25 people.

Table continued next page.

What Are the Other Provisions?	
Systems collecting fewer than five ROUTINE samples per month...	Must have a sanitary survey every 5 years (or every 10 years if it is a noncommunity water system using protected and disinfected groundwater).**
Systems using surface water or GWUDI and meeting filtration avoidance criteria...	Must collect and have analyzed one coliform sample each day the turbidity of the source water exceeds 1 ntu. This sample must be collected from a tap near the first service connection.
**As per the IESWTR, states must conduct sanitary surveys for community surface water and GWUDI systems in this category every 3 years	

How Is Compliance Determined?

- Compliance is based on the presence or absence of total coliforms.
- Compliance is determined each calendar month the system serves water to the public (or each calendar month that sampling occurs for systems on reduced monitoring).
- The results of ROUTINE and REPEAT samples are used to calculate compliance.

A Monthly MCL Violation Is Triggered If:	
A system collecting fewer than 40 samples per month...	Has more than one ROUTINE/REPEAT sample per month that is total coliform-positive.
A system collecting at least 40 samples per month...	Has more than 5.0 percent of the ROUTINE/REPEAT samples in a month total coliform-positive.

An Acute MCL Violation Is Triggered If:	
Any public water system...	Has any fecal coliform—or *E. coli*-positive REPEAT sample or has a fecal coliform—or *E. coli*-positive ROUTINE sample followed by a total coliform-positive REPEAT sample.

What Are the Public Notification and Reporting Requirements?	
For a monthly MCL violation...	• The violation must be reported to the state no later than the end of the next business day after the system learns of the violation. • The public must be notified within 14 days.*
For an acute MCL violation...	• The violation must be reported to the state no later than the end of the next business day after the system learns of the violation. • The public must be notified within 72 hours.*
Systems with ROUTINE or REPEAT samples that are fecal coliform– or *E. coli*–positive...	Must notify the state by the end of the day they are notified of the result or by the end of the next business day if the state office is already closed.
*The revised Public Notification Rule will extend the period allowed for public notice of monthly violations for 30 days and shorten the period for acute violations to 24 hours. These revisions are effective for all systems by May 6, 2002, and are detailed in 40 CFR Subpart Q.	

COMPREHENSIVE SURFACE WATER TREATMENT RULES (EXCEPT LT2ESWTR)—SYSTEMS USING SLOW SAND, DIATOMACEOUS EARTH, OR ALTERNATIVE FILTRATION

Overview	
Title	Surface Water Treatment Rule (SWTR): 40 CFR 141.70–141.75 Interim Enhanced Surface Water Treatment Rule (IESWTR): 40 CFR 141.170–141.175 Long-Term 1 Enhanced Surface Water Treatment Rule (LT1ESWTR): 40 CFR 141.500–141.571
Purpose	Improve public health protection through the control of microbial contaminants, particularly viruses, *Giardia*, and *Cryptosporidium.*
General Description	The Surface Water Treatment Rules: • Apply to all PWSs using surface water or GWUDI, otherwise known as "Subpart H Description systems." • Require *all* Subpart H systems to disinfect. • Require Subpart H systems to filter unless specific filter avoidance criteria are met. • Apply a treatment technique requirement for control of microbials.

Requirements				
This table shows how the requirements for the IESWTR and LT1ESWTR build on the existing requirements established in the original SWTR.				
Applicability: PWSs that use surface water or groundwater under the direct influence of surface water (Subpart H) that practice slow sand, diatomaceous earth, or alternative filtration.		Final Rule Dates		
		SWTR 1989	IESWTR 1998	LT1ESWTR 2002
Population Served	≥10,000	✓	✓	
	<10,000	✓	N/A (except for sanitary survey provisions)	✓
Regulated Pathogens	99.99 percent (4-log) removal/inactivation of viruses	✓	Regulated under SWTR	Regulated under SWTR
	99.9 percent (3-log) removal/inactivation of *Giardia lamblia*	✓	Regulated under SWTR	Regulated under SWTR
	99 percent (2-log) removal of *Cryptosporidium*		✓	✓
Residual Disinfectant Requirements	Entrance to distribution system (≥0.2 mg/L)	✓	Regulated under SWTR	Regulated under SWTR
	Detectable in the distribution system	✓	Regulated under SWTR	Regulated under SWTR

Table continued next page.

Requirements (continued)				
Applicability: PWSs that use surface water or groundwater under the direct influence of surface water (Subpart H) that practice slow sand, diatomaceous earth, or alternative filtration.		Final Rule Dates		
		SWTR 1989	IESWTR 1998	LT1ESWTR 2002
Turbidity Performance Standards	Combined filter effluent—slow sand and diatomaceous earth	✓	Regulated under SWTR	Regulated under SWTR
	Combined filter effluent—alternative	✓	✓	✓
Disinfection Profiling and Benchmarking	Systems must profile inactivation levels and generate benchmark, if required		✓	✓
Sanitary Surveys (state requirement)	Community water system: Every 3 years Noncommunity water system: Every 5 years		✓	Regulated under IESWTR
Covered Finished Reservoirs/Water Storage Facilities (new construction only)			✓	✓
Operated by Qualified Personnel as Specified by State		✓	Regulated under SWTR	Regulated under SWTR

Turbidity

Turbidity is measured as CFE for slow sand, diatomaceous earth, and alternative filtration. The CFE 95th percentile value and CFE maximum value for slow sand and diatomaceous earth were not lowered in the IESWTR and LT1ESWTR because these filtration technologies are assumed to provide 2-log *Cryptosporidium* removal with the turbidity limits established by the SWTR. Alternative filtration technologies (defined as filtration technologies other than conventional, direct, slow sand, or diatomaceous earth) must demonstrate to the state that filtration and/or disinfection achieve 3-log *Giardia* and 4-log virus removal and/or inactivation. The IESWTR and LT1ESWTR also require alternative filtration technologies to demonstrate 2-log *Cryptosporidium* removal.

Turbidity: Monitoring and Reporting Requirements

Turbidity Type and Reporting Requirements *(reports due by the tenth day of the following month the system serves water to the public)*		Monitoring/ Recording Frequency*	SWTR as of June 29, 1993	IESWTR ≥10,000 Persons as of Jan. 1, 2002	LT1ESWTR <10,000 Persons as of Jan. 1, 2005
Slow Sand and Diatomaceous Earth	CFE 95%	At least every 4 hours	≤1 ntu	Regulated under SWTR	Regulated under SWTR
	CFE max.	At least every 4 hours	5 ntu	Regulated under SWTR	Regulated under SWTR
Alternative •Membranes •Cartridges •Other	CFE 95%	At least every 4 hours	≤1 ntu	Established by state	Established by state (not to exceed 1 ntu)
	CFE max.	At least every 4 hours	5 ntu	Established by state	Established by state (not to exceed 5 ntu)

*Monitoring frequency may be reduced by the state to once per day for systems using slow sand or alternative filtration. Monitoring frequency may be reduced by the state to once per day for systems serving 500 or fewer people regardless of type of filtration used.

CFE Turbidity: Reporting Requirements

Report to State:	SWTR Measurements	IESWTR Measurements	LT1ESWTR Measurements*
Within 10 days after the end of the month:	Total number of monthly measurements	Total number of monthly measurements	Total number of monthly measurements
	Number and percent less than or equal to designated 95th percentile turbidity limits	Number and percent less than or equal to designated 95th percentile turbidity limits	Number and percent less than or equal to designated 95th percentile turbidity limits
	Date and value exceeding 5 ntu	Date and value exceeding 5 ntu for slow sand and diatomaceous earth or maximum level set by state for alternative filtration	Date and value exceeding 5 ntu for slow sand and diatomaceous earth or maximum level set by state for alternative filtration
Within 24 hours:	Exceedances of 5 ntu for CFE	Exceedances of 5 ntu for slow sand and diatomaceous earth or maximum CFE level set by state for alternative filtration	Exceedances of 5 ntu for slow sand and diatomaceous earth or maximum CFE level set by state for alternative filtration

*Systems serving fewer than 10,000 people must begin complying with these requirements beginning Jan. 1, 2005.

Disinfection

Disinfection must be sufficient to ensure that the system's total treatment process (disinfection plus filtration) achieves at least:

- 99.9% (3-log) inactivation and/or removal of *Giardia lamblia*
- 99.99% (4-log) inactivation and/or removal of viruses.

Cryptosporidium must be removed by filtration and no inactivation credits are currently given for disinfection. Systems must also comply with the MRDL requirements specified in the Stage 1 D/DBPR.

Residual Disinfectant: Monitoring and Reporting Requirements

Location	Concentration	Monitoring Frequency	Reporting (reports due on the 10th day of the following month)
Entry to distribution system	Residual disinfectant concentration cannot be <0.2 mg/L for more than 4 hours	Continuous, but states may allow systems serving 3,300 or fewer persons to take grab samples from one to four times per day, depending on system size	Lowest daily value for each day, the date and duration when residual disinfectant was <0.2 mg/L, and when state was notified of events where residual disinfectant was <0.2 mg/L
Distribution system—same location as total coliform sample location(s)	Residual disinfectant concentration cannot be undetectable in more than 5 percent of samples in a month, for any two consecutive months; HPC ≤500/mL is deemed to have detectable residual disinfectant	Same time as total coliform samples	Number of residual disinfectant or HPC measurements taken in the month resulting in no more than 5 percent of the measurements as being undetectable in any two consecutive months

Disinfection Profiling and Benchmarking

A *disinfection profile* is the graphical representation of a system's microbial inactivation over 12 consecutive months.

A *disinfection benchmark* is the lowest monthly average microbial inactivation value. The disinfection benchmark is used as a baseline of inactivation when considering changes in the disinfection process.

Disinfection Profiling and Benchmarking Requirements Under IESWTR and LT1ESWTR

The purpose of disinfection profiling and benchmarking is to allow systems and states to assess whether a change in disinfection practices creates a microbial risk. Systems should develop a disinfection profile that reflects *Giardia lamblia* inactivation (systems using ozone or chloramines must also calculate inactivation of viruses), calculate a benchmark (lowest monthly inactivation) based on the profile, and consult with the state prior to making a significant change to disinfection practices.

Requirement	IESWTR	LT1ESWTR
Affected Systems:	Community; nontransient, noncommunity; *and transient* systems	Community and nontransient, noncommunity systems only
Begin Profiling by:	Apr. 1, 2000	• July 1, 2003, for systems serving 500–9,999 people • Jan. 1, 2004, for systems serving fewer than 500 people
Frequency and Duration:	Daily monitoring for 12 consecutive calendar months to determine the total logs of *Giardia lamblia* inactivation (and viruses, if necessary) for each day in operation	Weekly inactivation of *Giardia lamblia* (and viruses, if necessary) on the same calendar day each week over 12 consecutive months
States May Waive Disinfection Profiling Requirements if:	Total trihalomethane annual average is <0.064 mg/L *and* sum of 5 haloacetic acid annual average is <0.048 mg/L: • Collected during the same period • Annual average is arithmetic average of the quarterly averages of four consecutive quarters of monitoring • At least 25 percent of samples at the maximum residence time in the distribution system • Remaining 75 percent of samples at representative locations in the distribution system	One total trihalomethane sample is <0.064 mg/L and one sum of 5 haloacetic acids sample is <0.048 mg/L: • Collected during the month of warmest water temperature; *and* • At the maximum residence time in the distribution system • Samples must have been collected after Jan. 1, 1998
Disinfection Benchmark Must be Calculated if:	Systems required to develop a disinfection profile and are considering any of the following: • Changes to the point of disinfection • Changes to the disinfectants used • Changes to the disinfection process • Any other modification identified by the state Systems must consult the state prior to making any modifications to disinfection practices.	Same as IESWTR, and systems must obtain state approval prior to making any modifications to disinfection practices

COMPREHENSIVE SURFACE WATER TREATMENT RULES (EXCEPT LT2ESWTR)—SYSTEMS USING CONVENTIONAL OR DIRECT FILTRATION

Overview	
Title	Surface Water Treatment Rule (SWTR): 40 CFR 141.70–141.75 Interim Enhanced Surface Water Treatment Rule (IESWTR): 40 CFR 141.170–141.175 Filter Backwash Recycling Rule (FBRR): 40 CFR 141.76 Long-Term 1 Enhanced Surface Water Treatment Rule (LT1ESWTR): 40 CFR 141.500–141.571
Purpose	Improve public health protection through the control of microbial contaminants, particularly viruses, *Giardia*, and *Cryptosporidium*.
General Description	The Surface Water Treatment Rules: • Apply to all PWSs using surface water or GWUDI, otherwise known as "Subpart H systems." • Require *all* Subpart H systems to disinfect. • Require Subpart H systems to filter unless specific filter avoidance criteria are met. • Require individual filter monitoring and establish CFE limits. • Apply a treatment technique requirement for control of microbials.

Requirements

This table shows how the requirements for the IESWTR and LT1ESWTR build on the existing requirements established in the original SWTR.

Applicability: PWSs that use surface water or groundwater under the direct influence of surface water (Subpart H) that practice conventional or direct filtration.		Final Rule Dates			
		SWTR 1989	IESWTR 1998	LT1ESWTR 2002	FBRR 2001
Population Served	≥10,000	✓	✓		✓
	<10,000	✓	N/A (except for sanitary survey provisions)	✓	✓
Regulated Pathogens	99.99 percent (4-log) removal/inactivation of viruses	✓	Regulated under SWTR	Regulated under SWTR	Regulated under SWTR
	99.9 percent (3-log) removal/inactivation of *Giardia lamblia*	✓	Regulated under SWTR	Regulated under SWTR	Regulated under SWTR
	99 percent (2-log) removal of *Cryptosporidium*		✓	✓	Regulated under IESWTR and LT1ESWTR

Table continued next page.

Requirements (continued)						
Applicability: PWSs that use surface water or groundwater under the direct influence of surface water (Subpart H) that practice conventional or direct filtration.		Final Rule Dates				
		SWTR 1989	IESWTR 1998	LT1ESWTR 2002	FBRR 2001
Residual Disinfectant Requirements	Entrance to distribution system (≥0.2 mg/L)	✓	Regulated under SWTR	Regulated under SWTR	
	Detectable in the distribution system	✓	Regulated under SWTR	Regulated under SWTR	
Turbidity Performance Standards	Combined filter effluent	✓	✓	✓	
	Individual filter effluent		✓	✓	
Disinfection Profiling and Benchmarking	Systems must profile inactivation levels and generate benchmark, if required		✓	✓	
Sanitary Surveys (state requirement)	Community water system: Every 3 years Noncommunity water system: Every 5 years		✓	Regulated under IESWTR	
Covered Finished Reservoirs/Water Storage Facilities (new construction only)			✓	✓	
Operated by Qualified Personnel as Specified by State		✓	Regulated under SWTR	Regulated under SWTR	Regulated under SWTR

Turbidity

There are two ways turbidity is measured: *combined filter effluent* and *individual filter effluent.*

Turbidity: Monitoring and Reporting Requirements

Turbidity Reporting Requirements *(reports due by the tenth day of the following month the system serves water to the public)*	Monitoring/ Recording Frequency	SWTR as of June 29, 1993	IESWTR ≥10,000 People as of Jan. 1, 2002	LT1ESWTR <10,000 People as of Jan. 1, 2005
CFE 95th Percentile Value Report total number of CFE measurements and number and percentage of CFE measurements ≤95th percentile limit	At least every 4 hours*	≤0.5 ntu	≤0.3 ntu	≤0.3 ntu
CFE Maximum Value Report date and value of any CFE measurement that exceeded CFE maximum limit	At least every 4 hours*	5 ntu	1 ntu	1 ntu
		Contact state within 24 hours	Contact state within 24 hours	Contact state within 24 hours
IFE Monitoring Report IFE monitoring conducted and any follow-up actions	Monitor continuously every 15 minutes	None	Monitor—exceedances require follow-up action	Monitor—exceedances require follow-up action: systems with two or fewer filters may monitor CFE continuously in lieu of IFE

*Monitoring frequency may be reduced by the state to once per day for systems serving 500 or fewer people.

IFE Follow-Up and Reporting Requirements

Condition	IESWTR (≥10,000)			LT1ESWTR (<10,000)*		
	Action	Report	By	Action	Report	By
Two consecutive recordings >0.5 ntu taken 15 minutes apart at the end of the first 4 hours of continuous filter operation after backwash/offline	Produce filter profile within 7 days (if cause not known)	• Filter number • Turbidity value • Date • Cause (if known) *or* report profile was produced	10th of the following month			
Two consecutive recordings >1.0 ntu taken 15 minutes apart	Produce filter profile within 7 days (if cause not known)	• Filter number • Turbidity value • Date • Cause (if known) *or* report profile was produced	10th of the following month		• Filter number • Turbidity value • Date • Cause (if known)	10th of the following month
Two consecutive recordings >1.0 ntu taken 15 minutes apart at the same filter for 3 months in a row	Conduct filter self-assessment within 14 days	• Filter number • Turbidity value • Date • Report filter self-assessment produced	10th of the following month	Conduct a filter self-assessment within 14 days; systems with two filters that monitor CFE in lieu of IFE must do both filters	• Date filter self-assessment triggered and completed	10th of the following month (or within 14 days of filter self-assessment being triggered if triggered in last 4 days of the month)
Two consecutive recordings >2.0 ntu taken 15 minutes apart at the same filter for 2 months in a row	Arrange for CPE within 30 days and submit report within 90 days	• Filter number • Turbidity value • Date	10th of the following month	Arrange for CPE within 60 days and submit CPE report within 120 days	• Date CPE triggered	10th of the following month
		Submit CPE report	90 days after exceedance		Submit CPE report	120 days after exceedance

*Systems serving fewer than 10,000 people must begin complying with these requirements beginning Jan. 1, 2005.

IFE performance is measured in systems using conventional or direct filtration. The performance of each individual filter is critical to controlling pathogen breakthrough. The CFE turbidity results may mask the performance of an individual filter because the individual filter may have a turbidity spike of a short duration not detected by 4-hour CFE readings.

The IESWTR and LT1ESWTR created more stringent CFE turbidity standards and established a new IFE turbidity monitoring requirement to address *Cryptosporidium*. These new turbidity standards assure conventional and direct filtration systems will be able to provide 2-log *Cryptosporidium* removal.

Disinfection

Disinfection must be sufficient to ensure that the system's total treatment process (disinfection plus filtration) achieves at least:

- 99.9% (3-log) inactivation and/or removal of *Giardia lamblia*
- 99.99% (4-log) inactivation and/or removal of viruses.

Cryptosporidium must be removed by filtration and no inactivation credits are currently given for disinfection. Systems must also comply with the MRDL requirements specified in the Stage 1 D/DBPR.

Residual Disinfectant: Monitoring and Reporting Requirements

Location	Concentration	Monitoring Frequency	Reporting (reports due on the 10th day of the following month)
Entry to distribution system	Residual disinfectant concentration cannot be <0.2 mg/L for more than 4 hours	Continuous, but states may allow systems serving 3,300 or fewer persons to take grab samples from one to four times per day, depending on system size	Lowest daily value for each day, the date and duration when residual disinfectant was <0.2 mg/L, and when state was notified of events where residual disinfectant was <0.2 mg/L
Distribution system—same location as total coliform sample location(s)	Residual disinfectant concentration cannot be undetectable in greater than 5 percent of samples in a month; for any two consecutive months; HPC ≤500/mL is deemed to have detectable residual disinfectant	Same time as total coliform samples	Number of residual disinfectant or HPC measurements taken in the month resulting in no more than 5 percent of the measurements as being undetectable in any two consecutive months

Disinfection Profiling and Benchmarking

A *disinfection profile* is the graphical representation of a system's microbial inactivation over 12 consecutive months.

A *disinfection benchmark* is the lowest monthly average microbial inactivation value. The disinfection benchmark is used as a baseline of inactivation when considering changes in the disinfection process.

Disinfection Profiling and Benchmarking Requirements Under IESWTR and LT1ESWTR

The purpose of disinfection profiling and benchmarking is to allow systems and states to assess whether a change in disinfection practices creates a microbial risk. Systems should develop a disinfection profile that reflects *Giardia lamblia* inactivation (systems using ozone or chloramines must also calculate inactivation of viruses), calculate a benchmark (lowest monthly inactivation) based on the profile, and consult with the state prior to making a significant change to disinfection practices.

Requirement	IESWTR	LT1ESWTR
Affected Systems:	Community; nontransient, noncommunity; *and transient* systems	Community and nontransient, noncommunity systems only
Begin Profiling by:	Apr. 1, 2000	• July 1, 2003, for systems serving 500–9,999 people • Jan. 1, 2004, for systems serving fewer than 500 people
Frequency and Duration:	Daily monitoring for 12 consecutive calendar months to determine the total logs of *Giardia lamblia* inactivation (and viruses, if necessary) for each day in operation	Weekly inactivation of *Giardia lamblia* (and viruses, if necessary) on the same calendar day each week over 12 consecutive months
States May Waive Disinfection Profiling Requirements if:	Total trihalomethanes annual average is <0.064 mg/L *and* sum of 5 haloacetic acids annual average is <0.048 mg/L: • Collected during the same period • Annual average is arithmetic average of the quarterly averages of four consecutive quarters of monitoring • At least 25 percent of samples at the maximum residence time in the distribution system • Remaining 75 percent of samples at representative locations in the distribution system	One total trihalomethanes sample is <0.064 mg/L *and* one sum of 5 haloacetic acids sample is <0.048 mg/L: • Collected during the month of warmest water temperature; and • At the maximum residence time in the distribution system Samples must have been collected after Jan. 1, 1998
Disinfection Benchmark Must Be Calculated if:	Systems required to develop a disinfection profile and are considering any of the following: • Changes to the point of disinfection • Changes to the disinfectant(s) used • Changes to the disinfection process • Any other modification identified by the state Systems must consult the state prior to making any modifications to disinfection practices	Same as IESWTR, and systems must obtain state approval prior to making any modifications to disinfection practices

Filter Backwash Recycling Rule

The FBRR applies to PWSs that use surface water or groundwater under the direct influence of surface water, practice conventional or direct filtration, and recycle spent filter backwash, thickener supernatant, or liquids from dewatering processes. The FBRR requires systems that recycle to return specific recycle flows through all processes of the system's existing conventional or direct filtration system or at an alternate location approved by the state. The FBRR was developed to improve public health protection by assessing and changing, where needed, recycle practices for improved contaminant control, particularly microbial contaminants. Systems were required to submit recycle notification to the state by Dec. 8, 2003.

Filter Backwash Critical Deadlines and Requirements	
June 8, 2004	• Return recycle flows through the processes of a system's existing conventional or direct filtration system or an alternate recycle location approved by the state (a 2-year extension is available for systems making capital improvements to modify the recycle return location) • Collect recycle flow information and retain on file
June 8, 2006	Complete all capital improvements associated with relocating recycle return location (if necessary)

COMPREHENSIVE SURFACE WATER TREATMENT RULES (EXCEPT LT2ESWTR)—UNFILTERED SYSTEMS

Overview	
Title	Surface Water Treatment Rule (SWTR): 40 CFR 141.70–141.75 Interim Enhanced Surface Water Treatment Rule (IESWTR): 40 CFR 141.170–141.175 Long-Term 1 Enhanced Surface Water Treatment Rule (LT1ESWTR): 40 CFR 141.500–141.571
Purpose	Improve public health protection through the control of microbial contaminants, particularly viruses, *Giardia,* and *Cryptosporidium.*
General Description	The Surface Water Treatment Rules: • Apply to all PWSs using surface water or GWUDI, otherwise known as "Subpart H systems." • Require *all* Subpart H systems to disinfect. • Require Subpart H systems to filter unless specific filter avoidance criteria are met. • Require unfiltered systems to perform source water monitoring and meet site-specific conditions for control of microbials.

Requirements

This table shows how the requirements for the IESWTR and LT1ESWTR build on the existing requirements established in the original SWTR.

Applicability: PWSs that use surface water or groundwater under the direct influence of surface water (Subpart H) that do not provide filtration.		Final Rule Dates		
		SWTR 1989	IESWTR 1998	LT1ESWTR 2002
Population Served	≥10,000	✓	✓	
	<10,000	✓	N/A (except for sanitary survey provisions)	✓
Regulated Pathogens	99.99 percent (4-log) inactivation of viruses	✓	Regulated under SWTR	Regulated under SWTR
	99.9 percent (3-log) inactivation of *Giardia lamblia*	✓	Regulated under SWTR	Regulated under SWTR
	99 percent (2-log) removal of *Cryptosporidium* (through watershed control)		✓	✓
Residual Disinfectant Requirements	Entrance to distribution system (≥0.2 mg/L)	✓	Regulated under SWTR	Regulated under SWTR
	Detectable in the distribution system	✓	Regulated under SWTR	Regulated under SWTR
Unfiltered System Requirements	Avoidance criteria	✓	✓	✓
Disinfection Profiling and Benchmarking	Systems must profile inactivation levels and generate benchmark, if required		✓	✓

Table continued next page.

Requirements (continued)				
This table shows how the requirements for the IESWTR and LT1ESWTR build on the existing requirements established in the original SWTR.				
Applicability: PWSs that use surface water or groundwater under the direct influence of surface water (Subpart H) that do not provide filtration.		Final Rule Dates		
		SWTR 1989	IESWTR 1998	LT1ESWTR 2002
Sanitary Surveys (state requirement)	Community water systems: • Every 3 years Noncommunity water systems: • Every 5 years		✓	Regulated under IESWTR
Covered Finished Reservoirs/Water Storage Facilities (new construction only)			✓	✓
Operated by Qualified Personnel as Specified by State		✓	Regulated under SWTR	Regulated under SWTR

Filtration Avoidance

Since Dec. 30, 1991, systems must meet source water quality and site-specific conditions to remain unfiltered. If any of the following criteria to avoid filtration are not met, systems must install filtration treatment within 18 months of the failure. The following table outlines the avoidance criteria established by the SWTR and later enhanced by the IESWTR and LT1ESWTR.

Filtration Avoidance Criteria			
		Requirement	Frequency
Source Water Quality Conditions	Microbial Quality	Monitor fecal coliform or total coliform density in representative samples of source water immediately prior to the first point of disinfectant application: • Fecal coliform density concentrations must be ≤20/100 mL; *or* • Total coliform density concentrations must be ≤100/100 mL Sample results must satisfy the criteria listed above in at least 90 percent of the measurements from previous 6 months	1 to 5 samples per week depending on system size and every day the turbidity of the source water exceeds 1 ntu
Source Water Quality Conditions	Turbidity	Prior to the first point of disinfectant application, turbidity levels cannot exceed 5 ntu	Performed on representative grab samples of source water every 4 hours (or more frequently)

Table continued next page.

Filtration Avoidance Criteria (continued)

		Requirement	Frequency
Site-Specific Conditions	Systems must:	Calculate total inactivation ratio daily and provide 3-log *Giardia lamblia* and 4-log virus inactivation daily (except any one day each month) in 11 of 12 previous months (on an ongoing basis)	Take daily measurements before or at the first customer at each residual disinfectant concentration sampling point: • Temperature • pH (if chlorine used) • Disinfectant contact time (at peak hourly flow) • Residual disinfectant concentration measurements (at peak hourly flow)
	System must comply with:	• MCL for total coliforms in 11 of 12 previous months (as per Total Coliform Rule) • Stage 1 Disinfectants/Disinfection By-products Rule requirements (as of Jan. 1, 2002, for systems serving ≥10,000 or Jan. 1, 2004, for systems serving <10,000)	
	Systems must have:	• Adequate entry point residual disinfectant concentration (see disinfection requirements) • Detectable residual disinfectant concentration in the distribution system (see disinfection requirements) • Redundant disinfection components or automatic shut-off whenever residual disinfectant concentration <0.2 mg/L • A watershed control program minimizing potential for contamination by *Giardia lamblia* cysts and viruses in source water; IESWTR and LT1ESWTR update this requirement by adding *Cryptosporidium* control measures • An annual on-site inspection by state or approved third party with reported findings • Not been identified as a source of a waterborne disease outbreak	

Disinfection

Disinfection must be sufficient to ensure that the system's total treatment process achieves at least:

- 99.9% (3-log) inactivation of *Giardia lamblia*
- 99.99% (4-log) inactivation of viruses.

Currently, *Cryptosporidium* must be controlled through the watershed control program and no inactivation credits are currently given for disinfection. Systems must also comply with the MRDL requirements specified in the Stage 1 D/DBPR.

Residual Disinfectant: Monitoring and Reporting Requirements			
Location	Concentration	Monitoring Frequency	Reporting (reports due on the 10th day of the following month)
Entry to distribution system	Residual disinfectant concentration cannot be <0.2 mg/L for more than 4 hours	Continuous, but states may allow systems serving 3,300 or fewer persons to take grab samples from one to four times per day, depending on system size	Lowest daily value for each day, the date and duration when residual disinfectant was <0.2 mg/L, and when state was notified of events where residual disinfectant was <0.2 mg/L
Distribution system—same location as total coliform sample location(s)	Residual disinfectant concentration cannot be undetectable in greater than 5 percent of samples in a month, for any two consecutive months Heterotrophic plate count (HPC) ≤500/mL is deemed to have detectable residual disinfectant	Same time as total coliform samples	Number of residual disinfectant or HPC measurements taken in the month resulting in no more than 5 percent of the measurements as being undetectable in any two consecutive months

System Reporting Requirements	
Report to State	What to Report
Within 10 days after the end of the month:	• Source water quality information (microbial quality and turbidity measurements) • In addition to the disinfection information above, systems must report the daily residual disinfectant concentration(s) and disinfectant contact time(s) used for calculating the CT value(s)
By October 10 each year:	• Report compliance with all watershed control program requirements • Report on the on-site inspection unless conducted by state in which the state must provide the system a copy of the report
Within 24 hours:	• Turbidity exceedances of 5 ntu and waterborne disease outbreaks
As soon as possible but no later than the end of the next business day:	• Instance where the residual disinfectant level entering the distribution system was less than 0.2 mg/L

Disinfection Profiling and Benchmarking

A *disinfection profile* is the graphical representation of a system's microbial inactivation over 12 consecutive months.

A *disinfection benchmark* is the lowest monthly average microbial inactivation value. The disinfection benchmark is used as a baseline of inactivation when considering changes in the disinfection process.

Disinfection Profiling and Benchmarking Requirements Under IESWTR and LT1ESWTR

The purpose of disinfection profiling and benchmarking is to allow systems and states to assess whether a change in disinfection practices creates a microbial risk. Systems should develop a disinfection profile that reflects *Giardia lamblia* inactivation (systems using ozone or chloramines must also calculate inactivation of viruses), calculate a benchmark (lowest monthly inactivation) based on the profile, and consult with the state prior to making a significant change to disinfection practices.

Requirement	IESWTR	LT1ESWTR
Affected Systems:	Community; nontransient, noncommunity; *and transient* systems	Community and nontransient, noncommunity systems only
Begin Profiling by:	Apr. 1, 2000	• July 1, 2003, for systems serving 500–9,999 people • Jan. 1, 2004, for systems serving fewer than 500 people
Frequency and Duration:	Daily monitoring for 12 consecutive calendar months to determine the total logs of *Giardia lamblia* inactivation (and viruses, if necessary) for each day in operation	Weekly inactivation of *Giardia lamblia* (and viruses, if necessary) on the same calendar day each week over 12 consecutive months
States May Waive Disinfection Profiling Requirements if:	Total trihalomethanes annual average is <0.064 mg/L *and* sum of 5 haloacetic acid annual average is <0.048 mg/L: • Collected during the same period • Annual average is arithmetic average of the quarterly averages of four consecutive quarters of monitoring • At least 25 percent of samples at the maximum residence time in the distribution system • Remaining 75 percent of samples at representative locations in the distribution system	One total trihalomethanes sample <0.064 mg/L *and* one sum of 5 haloacetic acid sample is <0.048 mg/L: • Collected during the month of warmest water temperature; *and* • At the maximum residence time in the distribution system Samples must have been collected after Jan. 1, 1998
Disinfection Benchmark Must Be Calculated if:	Systems required to develop a disinfection profile and are considering any of the following: • Changes to the point of disinfection • Changes to the disinfectant(s) used • Changes to the disinfection process • Any other modification identified by the state Systems must consult the state prior to making any modifications to disinfection practices	Same as IESWTR, and systems must obtain state approval prior to making any modifications to disinfection practices

INTERIM ENHANCED SURFACE WATER TREATMENT RULE

Overview	
Title	Interim Enhanced Surface Water Treatment Rule (IESWTR): 63 FR 69478–69521, Dec. 16, 1998, Vol. 63, No. 241 Revisions to the IESWTR, the Stage 1 Disinfectants/Disinfection By-products Rule (Stage 1 D/DBPR), and revisions to State Primacy Requirements to Implement the Safe Drinking Water Act (SDWA) Amendments: 66 FR 3770, Jan. 16, 2001, Vol. 66, No. 29
Purpose	Improve public health control of microbial contaminants, particularly *Cryptosporidium.* Prevent significant increases in microbial risk that might otherwise occur when systems implement the Stage 1 D/DBPR.
General Description	Builds on treatment technique approach and requirements of the 1989 Surface Water Treatment Rule. Relies on existing technologies currently in use at water treatment plants.
Utilities Covered	Sanitary survey requirements apply to all public water systems using surface water or groundwater under the direct influence of surface water, regardless of size. All remaining requirements apply to public water systems that use surface water or groundwater under the direct influence of surface water and serve 10,000 or more people.

Major Provisions	
Regulated Contaminants	
Cryptosporidium	• MCLG of zero • 99 percent (2-log) physical removal for systems that filter • Include in watershed control program for unfiltered systems
Turbidity Performance Standards	Conventional and direct filtration combined filter effluent: • ≤0.3 ntu in at least 95 percent of measurements taken each month • Maximum level of 1 ntu
Turbidity Monitoring Requirement (conventional and direct filtration)	
Combined Filter Effluent	Performed every 4 hours to ensure compliance with turbidity performance standards
Individual Filter Effluent	Performed continuously (every 15 minutes) to assist treatment plant operators in understanding and assessing filter performance
Additional Requirements	
	• Disinfection profiling and benchmarking • Construction of new uncovered finished water storage facilities prohibited • Sanitary surveys, conducted by the state, for all surface water and groundwater under the direct influence of surface water systems regardless of size (every 3 years for community water systems and every 5 years for noncommunity water systems)

Profiling and Benchmarking

Public water systems must evaluate impacts on microbial risk before changing disinfection practices to ensure adequate protection is maintained. The three major steps are:

1. Determine if a public water system needs to profile based on TTHM and the sum of HAA5 levels (applicability monitoring).
2. Develop a disinfection profile that reflects daily *Giardia lamblia* inactivation for at least a year (systems using ozone or chloramines must also calculate inactivation of viruses).
3. Calculate a disinfection benchmark (lowest monthly inactivation) based on the profile and consult with the state prior to making a significant change to disinfection practices.

Critical Deadlines and Requirements

For Drinking Water Systems	
Feb. 16, 1999	Construction of uncovered finished water reservoirs is prohibited.
March 1999	Public water systems lacking ICR or other occurrence data begin four quarters of applicability monitoring for TTHM and HAA5 to determine if disinfection profiling is necessary.
Apr. 16, 1999	Systems that have four consecutive quarters of HAA5 occurrence data that meet the TTHM monitoring requirements must submit data to the state to determine if disinfection profiling is necessary.
Dec. 31, 1999	Public water systems with ICR data must submit it to states to determine if disinfection profiling is necessary.
Apr. 1, 2000	Public water systems must begin developing a disinfection profile if their annual average (based on four quarters of data) for TTHM is ≥0.064 mg/L or HAA5 is ≥0.048 mg/L.
Mar. 31, 2001	Disinfection profile must be complete.
Jan. 1, 2002	Surface water systems or groundwater under the direct influence of surface water systems serving 10,000 or more people must comply with all IESWTR provisions (e.g., turbidity standards, individual filter monitoring).
For States	
Dec. 16, 2000	States submit IESWTR primacy revision applications to USEPA (triggers interim primacy).
January 2002	States begin first round of sanitary surveys.
Dec. 16, 2002	Primacy extension deadline—all states with an extension must submit primacy revision applications to USEPA.
December 2004	States must complete first round of sanitary surveys for community water systems.
December 2006	States must complete first round of sanitary surveys for noncommunity water systems.

Public Health Benefits	
Implementation of the IESWTR will result in...	• Increased protection against gastrointestinal illnesses from *Cryptosporidium* and other pathogens through improvements in filtration • Reduced likelihood of endemic illness from *Cryptosporidium* by 110,000 to 463,000 cases annually • Reduced likelihood of outbreaks of cryptosporidiosis
Estimated impacts of the IESWTR include...	• National total annualized cost: $307 million • 92 percent of households will incur an increase of less than $1 per month • Less than 1 percent of households will incur an increase of more than $5 per month (about $8 per month)

LONG-TERM 1 ENHANCED SURFACE WATER TREATMENT RULE

Overview	
Title	Long-Term 1 Enhanced Surface Water Treatment Rule (LT1ESWTR): 67 FR 1812, Jan. 14, 2002, Vol. 67, No. 9
Purpose	Improve public health protection through the control of microbial contaminants, particularly *Cryptosporidium*. Prevent significant increases in microbial risk that might otherwise occur when systems implement the Stage 1 Disinfectants and Disinfection By-products Rule.
General Description	Builds on the requirements of the 1989 SWTR. Smaller system counterpart of the IESWTR.
Utilities Covered	Public water systems that use surface water or GWUDI and serve fewer than 10,000 people.

Major Provisions	
Control of *Cryptosporidium*	• The MCLG is set at zero • Filtered systems must physically remove 99 percent (2-log) of *Cryptosporidium* • Unfiltered systems must update their watershed control programs to minimize the potential for contamination by *Cryptosporidium* oocysts • *Cryptosporidium* is included as an indicator of GWUDI
CFE Turbidity Performance Standards	*Specific CFE turbidity requirements depend on the type of filtration used by the system.* <u>Conventional and Direct Filtration</u> • ≤0.3 ntu in at least 95 percent of measurements taken each month • Maximum level of turbidity: 1 ntu <u>Slow Sand and DE Filtration</u> • Continue to meet CFE turbidity limits specified in the SWTR: — 1 ntu in at least 95 percent of measurements taken each month — Maximum level of turbidity: 5 ntu <u>Alternative Technologies (other than conventional, direct, slow sand, or DE)</u> • Turbidity levels are established by the state based on filter demonstration data submitted by the system — State-set limits must not exceed 1 ntu (in at least 95 percent of measurements) or 5 ntu (maximum)

Turbidity Monitoring Requirements

Combined Filter Effluent	Performed at least every 4 hours to ensure compliance with CFE turbidity performance standards*
IFE (for systems using conventional and direct filtration only)	*Because the CFE may meet regulatory requirements even though one filter is producing high-turbidity water, the IFE is measured to assist conventional and direct filtration treatment plant operators in understanding and assessing individual filter performance.* • Performed continuously (recorded at least every 15 minutes) • Systems with two or fewer filters may conduct continuous monitoring of CFE turbidity in place of individual filter effluent turbidity monitoring • Certain follow-up actions are required if the IFE turbidity (or CFE for systems with two filters) exceeds 1.0 ntu in two consecutive readings or more (i.e., additional reporting, filter self-assessments, and/or CPEs

*This frequency may be reduced by the state to once per day for systems using slow sand/alternative filtration or for systems serving 500 persons or fewer regardless of the type of filtration used.

Disinfection Profiling and Benchmarking

Community and nontransient, noncommunity public water systems must evaluate impacts on microbial risk before changing disinfection practices to ensure adequate microbial protection is maintained. This is accomplished through a process called disinfection profiling and benchmarking.

What are the disinfection profiling and benchmarking requirements?

- Systems must develop a disinfection profile, which is a graphical compilation of weekly inactivation of *Giardia lamblia,* taken on the same calendar day each week over 12 consecutive months. (Systems using chloramines, ozone, or chlorine dioxide for primary disinfection must also calculate inactivation of viruses). Results must be available for review by the state during sanitary surveys.
- A state may deem a profile unnecessary if the system has sample data collected after Jan. 1, 1998—during the month of warmest water temperature and at maximum residence time in the distribution system—indicating TTHM levels are below 0.064 mg/L *and* the sum of HAA5 levels are below 0.048 mg/L.
- Prior to making significant changes to disinfection practices, systems required to develop a profile must calculate a disinfection benchmark and consult with the state. The benchmark is the calculation of the lowest monthly average of inactivation based on the disinfection profile.

Additional Requirements

- Construction of new uncovered finished water reservoirs is prohibited.

Critical Deadlines and Requirements

For Drinking Water Systems	
March 15, 2002	Construction of uncovered finished reservoirs is prohibited
July 1, 2003	No later than this date, systems serving between 500 and 9,999 persons must report to the state: • Results of optional monitoring that show levels of TTHM <0.064 mg/L *and* HAA5 <0.048 mg/L or • System has started profiling
Jan. 1, 2004	No later than this date, systems serving fewer than 500 persons must report to the state: • Results of optional monitoring that show levels of TTHM <0.064 mg/L *and* HAA5 <0.048 mg/L or • System has started profiling
June 30, 2004	Systems serving between 500 and 9,999 persons must complete their disinfection profile unless the state has determined it is unnecessary.
Dec. 31, 2004	Systems serving fewer than 500 persons must complete their disinfection profile unless the state has determined it is unnecessary.
Jan. 14, 2005	Surface water systems or GWUDI systems serving fewer than 10,000 people must comply with the applicable LT1ESWTR provisions (e.g., turbidity standards, individual filter monitoring, *Cryptosporidium* removal requirements, updated watershed control requirements for unfiltered systems).
For States	
January 2002	As per the IESWTR, states begin first round of sanitary surveys (at least every 3 years for community water systems and every 5 years for noncommunity water systems).
Oct. 14, 2003	States are encouraged to submit final primacy applications to USEPA.
Jan. 14, 2004	Final primacy applications must be submitted to USEPA unless granted an extension.
December 2004	States must complete first round of sanitary surveys for community water systems (as per the IESWTR).
Jan. 14, 2006	Final primacy revision applications from states with approved 2-year extension agreements must be submitted to USEPA.
December 2006	States must complete first round of sanitary surveys for noncommunity water systems (as per the IESWTR).

Public Health Benefits

Implementation of the LT1ESWTR will result in...	• Increased protection against gastrointestinal illnesses from *Cryptosporidium* and other pathogens through improvements in filtration. • Reduced likelihood of endemic illness from *Cryptosporidium* by an estimated 12,000 to 41,000 cases annually. • Reduced likelihood of outbreaks of cryptosporidiosis.
Estimated impacts of the LT1ESWTR include...	• National total annualized cost: $39.5 million. • 90 percent of affected households will incur an increase of less than $1.25 per month. • One percent of affected households are likely to incur an increase of more than $10 per month.

LONG-TERM 2 ENHANCED SURFACE WATER TREATMENT RULE

This section provides essential information on the Long-Term 2 Enhanced Surface Water Treatment Rule (LT2ESWTR) that was promulgated on Jan. 5, 2006, particularly as it applies to public water systems that use surface water or groundwater under the direct influence (GWUDI) of surface water and serve 50,000 or more people. These systems have the earliest compliance deadlines, which are detailed here.

Implementation of the rule, which complements earlier surface water treatment rules, begins with initial source water monitoring by all affected systems, which must do so using approved laboratories that use approved methods. For purposes of implementing the rule on a staggered schedule, USEPA has placed utilities in one of four compliance "schedules" according to system size.

Also for this rule, USEPA has adopted a standardized national approach for regulating wholesale utilities that are part of an interconnected system of wholesale and consecutive systems. As defined in the companion Stage 2 Disinfectants/Disinfection By-products Rule, which was promulgated on Jan. 4, 2006,

- A combined distribution system is the interconnected distribution system consisting of the distribution systems of wholesale systems and of the consecutive systems that receive finished water.
- A wholesale system is a public water system that treats source water as necessary to produce finished water and then delivers some or all of that finished water to another public water system. Delivery may be through a direct connection or through the distribution system of one or more consecutive systems.
- A consecutive system is a public water system that receives some or all of its finished water from one or more wholesale systems. Delivery may be through a direct connection or through the distribution system of one or more consecutive systems.

The schedule groups are defined as follows:

- *Schedule 1* systems include those serving 100,000 or more people or wholesale systems that are part of a combined distribution system in which the largest system serves 100,000 or more people.
- *Schedule 2* systems include those serving 50,000 to 99,999 people or wholesale systems that are part of a combined distribution system in which the largest system serves 50,000 to 99,999 people.
- *Schedule 3* systems include those serving 10,000 to 49,999 people or wholesale systems that are part of a combined distribution system in which the largest system serves 10,000 to 49,999 people.
- *Schedule 4* systems include systems that serve fewer than 10,000 people and are not a wholesale system.

Those in Schedules 1, 2, and 3 will conduct 2 years of monitoring on staggered schedules. Filtered systems in these three schedules must sample for *Cryptosporidium*, *E. coli*, and turbidity at least monthly for 24 months, and unfiltered systems must sample for *Cryptosporidium* at least monthly for 24 months.

For Schedule 4 systems, filtered systems only must sample for *E. coli* (or an alternative indicator approved by the state) at least once every 2 weeks for 12 months unless they opt to skip

E. coli monitoring and comply with the later schedule for *Cryptosporidium* monitoring. All Schedule 4 unfiltered systems and filtered systems that conduct *E. coli* monitoring and exceed certain trigger levels must sample for *Cryptosporidium* at least twice monthly for 12 months or monthly for 24 months. Also, Schedule 4 systems using GWUDI must comply with source water monitoring requirements based on the *E. coli* level of the nearest surface water body.

For all source water monitoring, the rule requires water systems to submit monitoring plans electronically for review and approval. Once monitoring begins, laboratories are to report results electronically, and water systems will be able to review and approve the data. USEPA has established an Information Processing and Management Center with an electronic Data Collection and Tracking System to accept and manage such data.

As described in this section, filtered systems will use their source water monitoring results to determine which of four "risk bins" each plant belongs in, with most expected to wind up in Bin 1, which requires no additional treatment. Systems that are placed in the other bins must then achieve the required additional *Cryptosporidium* treatment by selecting from various treatment and management options in a "toolbox" of options associated with each bin. Small systems that are not required to monitor for *Cryptosporidium* will be automatically placed in Bin 1.

The toolbox specifies treatment credits for each option, which include two source protection and management options, three prefiltration options, three filter performance options, five additional filtration options, and three inactivation options. One of the hallmarks of the LT2ESWTR is its recognition of the efficacy of UV light to inactivate *Cryptosporidium*. USEPA expects UV, a relatively affordable technology, will be installed in approximately 1,000 surface water treatment plants to comply with the rule. The rule is also expected to lead approximately 1,500 mostly small systems to install bag filters to provide the necessary additional treatment for *Cryptosporidium*.

All filtered and unfiltered systems that monitor for *Cryptosporidium* must report their bin classification to state authorities no later than 6 months after completion of monitoring.

Filtered systems that wind up in any of the three risk bins requiring additional treatment must meet reporting requirements and deadlines for each treatment option selected and provide the bin-appropriate level of treatment within 3 years following bin assignment, unless the state allows up to an additional 2 years for systems that must make capital improvements. Unfiltered systems must provide 2-log or 3-log inactivation of *Cryptosporidium* depending on their source water monitoring results, and they must do so using at least two of three technologies: chlorine dioxide, ozone, or UV.

The key LT2ESWTR compliance deadlines, by schedule group, are listed in the following table:

LT2ESWTR Key Compliance Deadlines by System Size Schedule

Schedule	Submit Sample Plan	Begin Initial Monitoring	Submit Bin Classification	Meet Additional Treatment Requirements*
Schedule 1 systems (serving ≥100,000)	July 2006	October 2006	March 2009	Mar. 1, 2012
Schedule 2 systems (serving 50,000–99,999)	January 2007	April 2007	October 2009	Sept. 30, 2012
Schedule 3 systems (serving 10,000–49,999)	January 2008	April 2008	October 2010	Sept. 30, 2013
Schedule 4 systems (serving <10,000) *E. coli* monitoring *Cryptosporidium* monitoring	 July 2008 January 2010	 October 2008 April 2010	 October 2010	 Sept. 30, 2014

*If *Cryptosporidium* monitoring results place utility in Bin 2, 3, or 4. States can extend the deadline up to an additional 24 months if capital improvements are required.

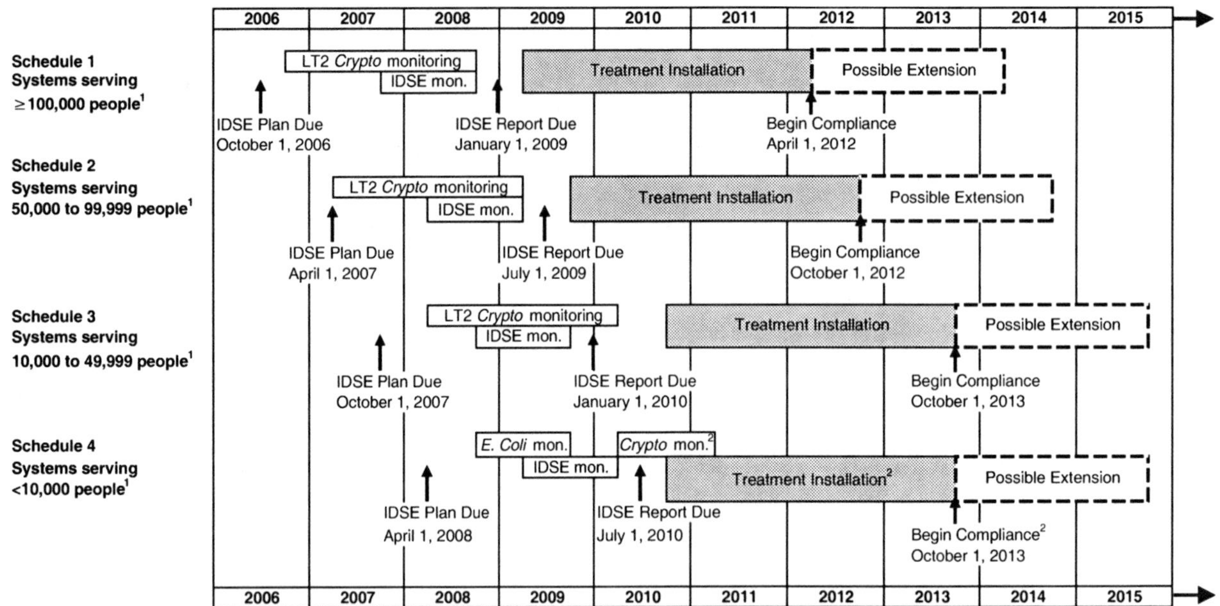

Figure LT2ESWTR-1 Stage 2 D/DBPR and LT2ESWTR compliance schedule

Also under the rule, all systems that store treated water in open reservoirs must, by April 2009, either cover the reservoirs or treat the discharges to inactivate 4-log virus, 3-log *Giardia lamblia,* and 2-log *Cryptosporidium.* Affected systems must also review their current level of microbial treatment before making a significant change in their disinfection practice to ensure they maintain protection against microbial pathogens as they take steps to reduce the formation of regulated DBPs under the Stage 2 D/DBPR.

Failure to meet requirements regarding sampling schedules, locations, laboratories, and methods are monitoring violations. Failure to report monitoring data and bin classification data and meet additional treatment requirements are treatment technique violations.

Finally, USEPA has prepared many compliance assistance resources, including guidance documents on source water monitoring, laboratory practices, UV and other toolbox treatment and management options, and simultaneous compliance with the Stage 2 D/DBPR and other regulations. These compliance assistance tools are online at www.epa.gov/ogwdw/disinfection/lt2/index.html.

Figure LT2ESWTR-1 provides a comprehensive compliance schedule for both the LT2ESWTR and the Stage 2 D/DBPR.

Overview	
Title	Long-Term 2 Enhanced Surface Water Treatment Rule (LT2ESWTR): 71 FR 654, Jan. 5, 2006, Vol. 71, No. 3
Purpose	Improve public health protection through the control of microbial contaminants by focusing on systems with elevated *Cryptosporidium* risk. Prevent significant increases in microbial risk that might otherwise occur when systems implement the Stage 2 Disinfectants and Disinfection By-products Rule (Stage 2 D/DBPR).
General Description	The LT2ESWTR requires systems to monitor their source water, calculate an average *Cryptosporidium* concentration, and use those results to determine if their source is vulnerable to contamination and may require additional treatment.
Utilities Covered	• Public water systems (PWSs) that use surface water or groundwater under the direct influence of surface water (GWUDI). • Schedule 1 systems include PWSs serving 100,000 or more people OR wholesale PWSs that are part of a combined distribution system in which the largest system serves 100,000 or more people.

Major Provisions	
Control of *Cryptosporidium*	
Source Water Monitoring	• Filtered and unfiltered systems must conduct 24 months of source water monitoring for *Cryptosporidium*. Filtered systems must also record source water *E. coli* and turbidity levels. Systems will be classified into one of four "bins" based on the results of their source water monitoring. These systems may also use previously collected data (i.e., grandfathered data), instead of monitoring. • Filtered systems providing at least 5.5-log of treatment for *Cryptosporidium* and unfiltered systems providing at least 3-log of treatment for *Cryptosporidium* and those systems that intend to install this level of treatment are not required to conduct source water monitoring.
Installation of Additional Treatment	• Filtered systems must provide additional treatment for *Cryptosporidium* based on their bin classification (average source water *Cryptosporidium* concentration), using treatment options from the "microbial toolbox." • Unfiltered systems must provide additional treatment for *Cryptosporidium* using chlorine dioxide, ozone, or UV.
Uncovered Finished Water Storage Facility	Systems with an uncovered finished water storage facility must either: • Cover the uncovered finished water storage facility; or • Treat the discharge to achieve inactivation and/or removal of at least 4-log for viruses, 3-log for *Giardia lamblia*, and 2-log for *Cryptosporidium*.
Disinfection Profiling and Benchmarking	
After completing the initial round of source water monitoring, any system that plans on making a significant change to their disinfection practices must: • Create disinfection profiles for *Giardia lamblia* and viruses; • Calculate a disinfection benchmark; and • Consult with the state prior to making a significant change in disinfection practice.	

Bin Classification for Filtered Systems

Cryptosporidium Concentration (oocysts/L)	Bin Classification	Additional *Cryptosporidium* Treatment Required: Conventional Filtration	Additional *Cryptosporidium* Treatment Required: Direct Filtration	Additional *Cryptosporidium* Treatment Required: Slow Sand or Diatomaceous Earth Filtration	Alternative Filtration
<0.075	Bin 1	No additional treatment required	No additional treatment required	No additional treatment required	No additional treatment required
0.075 to <1.0	Bin 2	1 log	1.5 log	1 log	(1)
1.0 to <3.0	Bin 3	2 log	2.5 log	2 log	(2)
≥3.0	Bin 4	2.5 log	3 log	2.5 log	(3)

(1) As determined by the state (or other primacy agency) such that the total removal/inactivation >4.0-log.
(2) As determined by the state (or other primacy agency) such that the total removal/inactivation >5.0-log.
(3) As determined by the state (or other primacy agency) such that the total removal/inactivation >5.5-log.

Inactivation Requirements for Unfiltered Systems

Cryptosporidium Concentration (oocysts/L)	Required *Cryptosporidium* Inactivation
≤0.01	2-log
>0.01	3-log

Critical Deadlines and Requirements

For Drinking Water Systems

Schedule 1 Deadlines	Schedule 2 Deadlines	Requirements
July 2006	January 2007	Systems must submit their: • Sampling schedule that specifies the dates of sample collection and location of sampling for initial source water monitoring to USEPA electronically.* • Notify USEPA or the state of the system's intent to submit results for grandfathering data. • Notify USEPA or the state of the system's intent to provide at least 5.5 log of treatment for *Cryptosporidium*.*
October 2006	April 2007	Systems must begin 24 months of source water monitoring.
Dec. 10, 2006	June 10, 2007	Systems submit results for first month of source water monitoring.
December 2006	June 2007	Systems must submit monitoring results for data that they want to have grandfathered.*
Apr. 1, 2008	Apr. 1, 2008	No later than this date, systems must notify the USEPA or the state of all uncovered treated water storage facilities.
September 2008	March 2009	Systems complete their initial round of source water monitoring.*

*These dates are based on the date your system will start collecting initial source water monitoring samples in October 2006.

Table continued next page.

Critical Deadlines and Requirements (continued)

For Drinking Water Systems		
Schedule 1 Deadlines	**Schedule 2 Deadlines**	**Requirements**
March 2009	October 2009	Filtered systems must report their initial bin classification to the USEPA or the state for approval.*
March 2009	October 2009	Unfiltered systems must report the mean of all *Cryptosporidium* sample results to the USEPA or the state.
Apr. 1, 2009	Apr. 1, 2009	No later than this date, uncovered finished water storage facilities must be covered, or the water must be treated before entry into the distribution system.
Mar. 31, 2012	Sept. 30, 2012	Systems must install and operate additional treatment in accordance with their bin classification.†
Jan. 1, 2015	July 1, 2015	Systems must submit their sampling schedule that specifies the dates of sample collection and location of sampling for second round of source water monitoring to the state.
Apr. 1, 2015	Oct. 1, 2015	• Systems are required to begin conducting a second round of source water monitoring. • Based on the results, systems must re-determine their bin classification and provide additional *Cryptosporidium* treatment, if necessary.
For States		
Schedule 1 Deadlines	**Schedule 2 Deadlines**	**Requirements**
January–June 2006	January–June 2006	States are encouraged to communicate with affected systems regarding LT2ESWTR requirements.
Apr. 1, 2007	Apr. 1, 2007	States are encouraged to communicate LT2ESWTR requirements related to treatment, uncovered finished water reservoirs, and disinfection profiling to affected systems.
Oct. 15, 2007	Oct. 15, 2007	States are encouraged to submit final primacy applications or extension requests to USEPA.
Jan. 15, 2008	Jan. 15, 2008	Final primacy applications must be submitted to USEPA, unless granted an extension.
June 30, 2008	Dec. 31, 2008	States should begin awarding *Cryptosporidium* treatment credit for primary treatments in place.
Jan. 15, 2010	Jan. 15, 2010	Final primacy revision applications from states with approved 2-year extensions agreements must be submitted to USEPA.
Dec. 31, 2012	June 30, 2013	States should award *Cryptosporidium* treatment credit for toolbox option implementation.

*These dates are based on the date your system will start collecting initial source water monitoring samples in October 2006.
†States may allow up to an additional 24 months for compliance for systems making capital improvements.

FILTER BACKWASH RECYCLING RULE

Overview	
Title	Filter Backwash Recycling Rule (FBRR): 66 FR 31086, June 8, 2001, Vol. 66, No. 111
Purpose	Improve public health protection by assessing and changing, where needed, recycle practices for improved contaminant control, particularly microbial contaminants.
General Description	The FBRR requires systems that recycle to return specific recycle flows through all processes of the system's existing conventional or direct filtration system or at an alternate location approved by the state.
Utilities Covered	Applies to public water systems that use surface water or groundwater under the direct influence of surface water, practice conventional or direct filtration, and recycle spent filter backwash, thickener supernatant, or liquids from dewatering processes.

Public Health Benefits	
Implementation of FBRR will result in...	• Reduction in risk of illness from microbial pathogens in drinking water, particularly *Cryptosporidium.*
Estimated impacts of the FBRR include...	• FBRR will apply to an estimated 4,650 systems serving 35 million Americans. • Fewer than 400 systems are expected to require capital improvements. • Annualized capital costs incurred by public water systems associated with recycle modifications are estimated to be $5.8 million. • Mean annual cost per household is estimated to be less than $1.70 for 99 percent of the affected households and between $1.70 and $100 for the remaining 1 percent of affected households.

Conventional and Direct Filtration

- Conventional filtration, as defined in 40 CFR 141.2, is a series of processes including coagulation, flocculation, sedimentation, and filtration resulting in substantial particulate removal. Conventional filtration is the most common type of filtration.
- Direct filtration, as defined in 40 CFR 141.2, is a series of processes including coagulation and filtration, but excluding sedimentation, and resulting in substantial particulate removal. Typically, direct filtration can be used only with high-quality raw water that has low levels of turbidity and suspended solids.

Recycle Flows

- *Spent filter backwash water*—A stream containingRecycle Flowsparticles that are dislodged from filter media when water is forced back through a filter (backwashed) to clean the filter.
- *Thickener supernatant*—A stream containing the decant from a sedimentation basin, clarifier, or other unit that is used to treat water, solids, or semi-solids from the primary treatment processes.
- *Liquids from dewatering processes*—A stream containing liquids generated from a unit used to concentrate solids for disposal.

Critical Deadlines and Requirements	
For Drinking Water Systems	
Dec. 8, 2003	Submit recycle notification to the state.
June 8, 2004	Return recycle flows through the processes of a system's existing conventional or direct filtration system or an alternate recycle location approved by the state (a 2-year extension is available for systems making capital improvements to modify recycle location). Collect recycle flow information and retain on file.
June 8, 2006	Complete all capital improvements associated with relocating recycle return location (if necessary).
For States	
June 8, 2003	States submit FBRR primacy revision application to USEPA (triggers interim primacy).
June 8, 2005	Primacy extension deadline—all states with an extension must submit primacy revision applications to USEPA.

What Does a Recycle Notification Include?

- Plant schematic showing origin of recycle flows, how recycle flows are conveyed, and return location of recycle flows.
- Typical recycle flows (gpm), highest observed plant flow experienced in the previous year (gpm), and design flow for the treatment plant (gpm).
- State-approved plant operating capacity (if applicable).

What Recycle Flow Information Does a System Need to Collect and Retain on File?

- Copy of recycle notification and information submitted to the state.
- List of all recycle flows and frequency with which they are returned.
- Average and maximum backwash flow rates through filters and average and maximum duration of filter backwash process (in minutes).
- Typical filter run length and written summary of how filter run length is determined.
- Type of treatment provided for recycle flows.
- Data on the physical dimension of the equalization and/or treatment units, typical and maximum hydraulic loading rates, types of treatment chemicals used, average dose, frequency of use, and frequency at which solids are removed, if applicable.

PROPOSED GROUND WATER RULE

In May 2000 USEPA proposed a rule (65 FR 30194) that specifies the appropriate use of disinfection in groundwater and addresses other components of groundwater systems to assure public health protection. The Ground Water Rule (GWR) establishes multiple barriers to protect against bacteria and viruses in drinking water from groundwater sources and will establish a targeted strategy to identify groundwater systems at high risk for fecal contamination. The GWR is scheduled to be issued as a final regulation in in late 2005 or early 2006.

Background

Although groundwater has historically been thought to be free of microbial contamination, recent research indicates that some groundwaters are a source of waterborne disease. Most cases of waterborne disease are characterized by gastrointestinal symptoms (diarrhea, vomiting, etc.) that are frequently self-limiting in healthy individuals and rarely require medical treatment. However, these same symptoms are much more serious and can be fatal for persons in susceptible subpopulations (such as young children, the elderly, and persons with compromised immune systems). In addition, research indicates that some viral pathogens found in groundwater are linked to long-term health effects (for example, adult onset diabetes, myocarditis). USEPA does not believe all groundwater systems are fecally contaminated; data indicate that only a small percentage of groundwater systems are contaminated. However, the severity of health impacts and the number of people potentially exposed to microbial pathogens in groundwater indicate that a regulatory response is warranted.

Presently, only surface water systems and systems using groundwater under the direct influence of surface water are required to disinfect their water supplies. The 1996 amendments to the Safe Drinking Water Act require USEPA to develop regulations that require disinfection of groundwater systems "as necessary" to protect the public health (§1412(b)(8)). The proposed GWR will specify when corrective action (including disinfection) is required to protect consumers who receive water from groundwater systems from bacteria and viruses.

This rule applies to public groundwater systems (systems that have at least 15 service connections or regularly serve at least 25 individuals daily at least 60 days out of the year). This rule also applies to any system that mixes surface water and groundwater if the groundwater is added directly to the distribution system and provided to consumers without treatment. The GWR does not apply to privately owned wells; however, USEPA recommends owners of private wells to test for coliform bacteria once each year.

While developing the proposal, USEPA consulted extensively with stakeholders. USEPA benefited from the stakeholders' participation in four public meetings across the country, and their comments are reflected in the proposed rule. USEPA also received valuable input from small entity representatives as part of the Small Business Regulatory Enforcement Fairness Act (SBREFA) panel. The GWR Small Business Advisory Panel met seven times from April 1998 to June 1998. Many of the panel's recommendations have been incorporated into the proposed rule.

In February 1999, USEPA posted an informal draft of the GWR preamble on the Internet. Approximately 300 copies were also mailed to participants of public meetings or to those who requested a copy. USEPA received valuable comments and stakeholder input from over 80 individuals representing states, trade associations, environmental interest groups, as well as individual stakeholders.

Public Comment on the Proposed Rule

USEPA took public comment on the proposed GWR for 60 days. The comment period closed Aug. 4, 2000. USEPA received over 250 comments. For more information, the general public can call the Safe Drinking Water Hotline at 800-426-4791. A fact sheet, the proposal, and additional information are also available at http://www.epa.gov/safewater/gwr/gwrfs.html.

What Requirements Are Proposed in the GWR?

- System sanitary surveys conducted by the state and identification of significant deficiencies;
- Hydrogeologic sensitivity assessments for undisinfected systems;
- Source water microbial monitoring by systems that do not disinfect and draw from hydrogeologically sensitive aquifers or have detected fecal indicators within the system's distribution system;
- Corrective action by any system with significant deficiencies or positive microbial samples indicating fecal contamination; and
- Compliance monitoring for systems that disinfect to ensure that they reliably achieve 4-log (99.99 percent) inactivation or removal of viruses.

The proposed requirements are discussed in greater detail below.

Sanitary Survey

Applies to:

- All groundwater systems

Frequency:

- Every 3 years for community water systems; 5 years for noncommunity water systems, consistent with the 1998 Interim Enhanced Surface Water Treatment Rule. (Community water systems serve the same populations year round, e.g., houses and apartment buildings. Noncommunity water systems do not serve the same people year round, e.g., schools, factories, office buildings, hospitals, gas stations, and campgrounds.)

Key components:

- State must perform each system's sanitary survey and address the eight elements from the joint USEPA and Association of State Drinking Water Administrators guidance.
- State must have authority to enforce corrective action requirements.
- State must provide a list of significant deficiencies (e.g., those that require corrective action) to the system within 30 days of identification of the deficiencies.

Hydrogeologic Sensitivity Assessment

Applies to:

- All groundwater systems that do not provide 4-log (99.99 percent) virus inactivation/removal.

Frequency:

- One-time assessment of sensitivity (within 6 years of the final rule's date of publication for community water systems and 8 years for noncommunity water systems). Sensitive systems must monitor monthly (see below).

Key components:

- State must conduct a one-time assessment of all systems that do not provide 4-log virus inactivation/removal to identify those systems located in sensitive aquifers.
- USEPA considers karst, gravel, or fractured bedrock aquifers to be "sensitive" to microbial contamination. States may waive source water monitoring for sensitive systems if there is a hydrogeologic barrier to fecal contamination.

Source Water Monitoring

Applies to:

- Groundwater systems that are sensitive or have contamination in their distribution system ("triggered monitoring") and do not treat to 4-log removal or inactivation of viruses.

Frequency:

- Monthly for sensitive systems; once for triggered monitoring.

Key Components:

- *Routine monitoring.* For systems determined by the state to be hydrogeologically sensitive, the system must conduct monthly source water monitoring for fecal indicators. Sampling frequency may be reduced after 12 negative samples.
- *Triggered monitoring.* If a total coliform-positive sample is found in the distribution system, the system must collect one source water sample and monitor for a fecal indicator.

Corrective Action

Applies to:

- Groundwater systems that have a significant deficiency or have detected a fecal indicator in their source water.

Frequency:

- Correct within 90 days or longer with a state-approved schedule.

Key components:

- *Significant deficiency or source water contamination.* If a groundwater system is notified of significant deficiencies by the state or notified of a positive source water sample, within 90 days it must correct the contamination problem by eliminating the contamination source, correct the significant deficiencies, provide an alternative source water or install a treatment process that reliably achieves 4-log removal or inactivation of viruses. A system may take longer than 90 days for corrective action with a state-approved plan. Systems must notify the state of completion of the corrective action or the state must confirm correction within 30 days after the 90-day period or scheduled correction date.

- *Treatment.* Systems providing treatment must monitor treatment to ensure at least 4-log virus inactivation and/or removal.

Compliance Monitoring

Applies to:

- All groundwater systems that notify states they disinfect to avoid source water monitoring and to systems that disinfect as a corrective action.

Frequency:

- Systems serving less than 3,300 must monitor disinfection treatment once daily, while systems serving 3,300 or more people must monitor their disinfection treatment continuously.

Key components:

- If monitoring shows the disinfection concentration to be below the required level, the system must restore the disinfection concentration within 4 hours or notify the state.

For general information, contact the Safe Drinking Water Hotline at (800) 426-4791. The hotline is open Monday through Friday, excluding federal holidays, from 9:00 a.m. to 5:30 p.m. Eastern time.

Summaries of Regulations to Control Chemical Contaminants

ARSENIC AND CLARIFICATIONS TO COMPLIANCE AND NEW SOURCE MONITORING RULE

Overview	
Title	Arsenic and Clarifications to Compliance and New Source Monitoring Rule: 66 FR 6976 (Jan. 22, 2001)
Purpose	To improve public health by reducing exposure to arsenic in drinking water.
General Description	• Changes the arsenic MCL from 50 ng/L to 10 pg/L; • Sets arsenic MCLG at 0; • Requires monitoring for new systems and new drinking water sources; • Clarifies the procedures for determining compliance with the MCLs for IOCs, SOCs, and VOCs.
Utilities Covered	All community water systems (CWSs) and nontransient, noncommunity water systems (NTNCWSs) must comply with the arsenic requirements. USEPA estimates that 3,024 CWSs and 1,080 NTNCWSs will have to install treatment to comply with the revised MCL.

Public Health Benefits	
Implementation of the Arsenic Rule will result in...	• Avoidance of 16 to 26 nonfatal bladder and lung cancers per year. • Avoidance of 21 to 30 fatal bladder and lung cancers per year. • Reduction in the frequency of noncarcinogenic diseases.

Critical Deadlines and Requirements	
Consumer Confidence Report*	
Report Due	**Report Requirements**
July 1, 2001	For the report covering calendar year 2000, systems that detect arsenic between 25 pg/L and 50 pg/L must include an educational statement in the CCRs.
July 1, 2002, and beyond	For reports covering calendar years 2001 and beyond, systems that detect arsenic between 5 pg/L and 10 pg/L must include an educational statement in the CCRs.
July 1, 2002–July 1, 2006	For reports covering calendar years 2001 to 2005, systems that detect arsenic between 10 pg/L and 50 pg/L must include a health effects statement in their CCRs.
July 1, 2007, and beyond	For reports covering calendar year 2006 and beyond, systems that are in violation of the arsenic MCL (10 pg/L) must include a health effects statement in their CCRs.
For Drinking Water Systems	
Jan. 22, 2004	All *new* systems/sources must collect initial monitoring samples for all IOCs, SOCs, and VOCs within a period and frequency determined by the state.

Table continued next page.

Critical Deadlines and Requirements (continued)	
For Drinking Water Systems (continued)	
Jan. 1, 2005	When allowed by the state, systems may grandfather data collected after this date.
Jan. 23, 2006	The new arsenic MCL of 10 pg/L becomes effective. All systems must begin monitoring or, when allowed by the state, submit data that meets grandfathering requirements.
Dec. 31, 2006	Surface water systems must complete initial monitoring or have a state-approved waiver.
Dec. 31, 2007	Groundwater systems must complete initial monitoring or have a state-approved waiver.
For States	
Spring 2001	USEPA meets and works with states to explain new rules and requirements and to initiate adoption and implementation activities.
Jan. 22, 2003	State primacy revision applications due.
Jan. 22, 2005	State primacy revision applications due from states that received 2-year extensions.
*For required educational and health effects statements, see 40 CFR 141.154.	

Compliance Determination (IOCs, VOCs, and SOCs)

1. Calculate compliance based on a running annual average at each sampling point.
2. Systems will not be in violation until 1 year of quarterly samples have been collected (unless fewer samples would cause the running annual average to be exceeded).
3. If a system does not collect all required samples, compliance will be based on the running annual average of the samples collected.

Monitoring Requirements for Total Arsenic*	
Initial Monitoring	
One sample after the effective date of the MCL (Jan. 23, 2006). Surface water systems must take annual samples. Groundwater systems must take one sample between 2005 and 2007.	
Reduced Monitoring	
If the initial monitoring result for arsenic is less than the MCL...	• Groundwater systems must collect one sample every 3 years. • Surface water systems must collect annual samples.
Increased Monitoring	
A system with a sampling point result above the MCL must collect quarterly samples at that sampling point, until the system is reliably and consistently below the MCL.	
*All samples must be collected at each entry point to the distribution system, unless otherwise specified by the state.	

LEAD AND COPPER RULE

Overview	
Title	**Lead and Copper Rule (LCR): 56 FR 26460–26564, June 7, 1991***
Purpose	Protect public health by minimizing lead (Pb) and copper (Cu) levels in drinking water, primarily by reducing water corrosivity. Pb and Cu enter drinking water mainly from corrosion of Pb- and Cu-containing plumbing materials.
General Description	Establishes AL of 0.015 mg/L for Pb and 1.3. mg/L for Cu based on 90th percentile level of tap water samples. An AL exceedance is not a violation but can trigger other requirements that include WQP monitoring, CCT, source water monitoring/treatment, public education, and LSLR.
Utilities Covered	All CWSs and NTNCWSs are subject to the LCR requirements.

*The June 1991 LCR was revised with the following Technical Amendments: 56 FR 32112, July 15, 1991; 57 FR 28785, June 29, 1992; 59 FR 33860, June 30, 1994; and the LCR Minor Revisions 65 FR 1950, Jan. 12, 2000.

Public Health Benefits	
Implementation of the LCR has resulted in...	• Reduction in risk of exposure to Pb that can cause damage to brain, red blood cells, and kidneys, especially for young children and pregnant women. • Reduction in risk of exposure to Cu that can cause stomach and intestinal distress, liver or kidney damage, and complications of Wilson's disease in genetically predisposed people.

Sampling Requirements
• First-draw samples must be collected by all CWSs and NTNCWSs at cold water taps in homes/buildings that are at high risk of Pb/Cu contamination as identified in 40 CFR 141.86(a).
• Number of sample sites is based on system size (see Table LCR-1).
• Systems must conduct monitoring every 6 months unless they qualify for reduced monitoring (see Table LCR-2).

Table LCR-1 Pb and Cu tap and WQP tap monitoring

Size Category	System Size	Number of Pb/Cu Tap Sample Sites		Number of WQP Tap Sampling Sites	
		Standard	Reduced	Standard	Reduced
Large	>100,000	100	50	25	10
	50,001–100,000	60	30	10	7
Medium	10,001–50,000	60	30	10	7
	3,301–10,000	40	20	3	3
Small	501–3,300	20	10	2	2
	101–500	10	5	1	1
	≤100	5	5	1	1

Table LCR-2 Criteria for reduced Pb/Cu tap monitoring*

Can Monitor...	If the System...
Annually	1. Serves ≤50,000 and is ≤ALs for two consecutive 6-month monitoring periods; *or* 2. Meets OWQP specifications for two consecutive 6-month monitoring periods.
Triennially	1. Serves 50,000 and is ≤ both ALs for three consecutive years of monitoring; *or* 2. Meets OWQP specifications for three consecutive years of monitoring; *or* 3. Has 90th percentile Pb levels ≤0.005 mg/L and 90th percentile Cu level ≤0.65 mg/L for two consecutive 6-month periods (i.e., accelerated reduced Pb/Cu tap monitoring; *or* 4. Meets the 40 CFR 141.81(b)(3) criteria.
Once every 9 years	Serves ≤3,300 and meets monitoring waiver criteria found at 40 CFR 141.86(g).

*Samples are collected at reduced number of sites (see Table LCR-1).

Treatment Technique and Sampling Requirements

Corrosion control treatment installation: All large systems (except systems that meet the requirements of 40 CFR 141.81(b)(2) or (3)) must install CCT. Medium and small systems that exceed either AL must install CCT.

Water quality parameter monitoring: All large systems are required to do WQP monitoring. Medium and small systems that exceed either AL are required to do WQP monitoring.

Treatment Technique and Sampling Requirements if the Action Level Is Exceeded
1. WQP Monitoring
•All systems serving >50,000 people and those systems serving ≤50,000 people if 90th percentile tap level is greater than either AL must take WQP samples during the same monitoring periods as Pb/Cu tap sample.
•Used to determine water corrosivity and, if needed, to help identify type of CCT to be installed and how CCT should be operated (i.e., establishes OWQP levels).
•WQPs include: pH, alkalinity, calcium, conductivity (initial WQP monitoring only), orthophosphate (if phosphate-based inhibitor is used), silica (if silicate-based inhibitor is used), and temperature (initial WQP monitoring only).
•Samples are collected within distribution system (i.e., WQP tap samples), with number of sites based on system size (see Table LCR-1) and at each EPTDS.
•Systems installing CCT must conduct follow-up monitoring for two consecutive 6-month periods—WQP tap monitoring is conducted semi-annually; EPTDS monitoring increases to every 2 weeks.
•After follow-up monitoring, state sets ranges of values for the OWQPs.
•Reduced WQP tap monitoring is available for systems in compliance with OWQPs; *reduced monitoring does not apply to EPTDS monitoring.*
•For systems ≤50,000, WQP monitoring is not required whenever 90th percentile tap levels are ≤ALs.
2. Public Education (PE)
•Only required if Pb AL is exceeded (no PE is required if only Cu AL is exceeded).
•Informs PWSs' customers about health effects, sources, and what can be done to reduce exposure.
•Includes billing inserts sent directly to customers, pamphlets, or brochures distributed to hospitals and other locations that provide services to pregnant women and children and, for some CWSs, newspaper notices and PSAs submitted to TV/radio stations.
•System must begin delivering materials within 50 days of Pb AL exceedance and continue every 6 months for PSAs and annually for all other forms of delivery for as long as it exceeds Pb AL.
•Different delivery methods and mandatory language for CWSs and NTNCWSs.
•Can discontinue delivery whenever less than or equal the Pb AL but must recommence if Pb AL subsequently exceeded.
•PE requirements are in addition to the public notification required in 40 CFR Subpart Q.
3. Source Water Monitoring and Treatment
•All systems that exceed Pb or Cu AL must collect source water samples to determine contribution from source water to total tap water Pb/Cu levels and make a SOWT recommendation within 6 months of the exceedance.
•One set of samples at each EPTDS is due within 6 months of first AL exceedance.
•If state requires SOWT, system has 24 months to install SOWT.
•After follow-up Pb/Cu tap and EPTDS monitoring, state sets maximum permissible limits for Pb and Cu in source.

Table continued next page.

Treatment Technique and Sampling Requirements if the Action Level Is Exceeded (continued)
4. Corrosion Control Treatment •Required for all large systems (except systems that meet the requirements of 40 CFR 141.81(b)(2) or (b)(3)) and medium/small systems that exceed either AL. The system shall recommend optimal CCT within 6 months. •Corrosion control study required for large systems. •If state requires study for medium or small systems, it must be completed within 18 months. •Once state determines type of CCT to be installed, PWS has 24 months to install CCT. •Systems installing CCT must conduct follow-up monitoring for two consecutive 6-month periods. •After follow-up Pb/Cu tap and WQP monitoring, state sets OWQPs. •Small and medium systems can stop CCT steps if both ALs for two consecutive 6-month monitoring periods.
If the system continued to exceed the AL after installing CCT and/or SOWT...
5. Lead Service Line (LSL) Monitoring •Two types of sampling associated with LSLR: — *Optional*: Monitoring from LSL to determine need to replace line. If all Pb samples from line ≤0.015 mg/L, LSL does not need to be replaced and counts as replaced line. — *Required*: Monitoring if entire LSL is not replaced to determine impact from "partial" LSLR. Sample is collected that is representative of water in service line that is partially replaced. •Monitoring only applies to system subject to LSLR. **6. Lead Service Line Replacement** •System must replace LSLs that contribute more than 0.015 mg/L to tap water levels. •Must replace 7 percent of LSL per year; state can require accelerated schedule. •If only a portion of a LSL is replaced, PWS must: — Notify customers at least 45 days prior to replacement about the potential for increased Pb levels; — Collect sample within 72 hours of replacement and mail/post results within 3 days of receipt of results. •Systems can discontinue LSLR whenever ≤Pb AL in tap water for two consecutive monitoring periods.

*Lead and Copper Rule—Clarification of Requirements for Collecting Samples and Calculating Compliance**

The following text is from a fact sheet USEPA prepared to accompany a formal guidance memorandum issued Nov. 23, 2004, to state and USEPA regional drinking water officials clarifying certain Lead and Copper Rule requirements for collecting samples and calculating compliance. The complete guidance memorandum, a product of the agency's ongoing national review of LCR implementation issues, is available online at USEPA's LCR review site, www.epa.gov/safewater/lcrmr/lead_review.html. The agency is expected to promulgate additional revisions to the rule in 2006 based on its review.

USEPA is releasing a guidance memorandum to reiterate and clarify specific regulatory requirements for lead in drinking water. The guidance memorandum is intended for USEPA regional and state staff who work in the drinking water program. The audience also includes water utilities who are subject to the regulations.

USEPA has been conducting a national review of implementation of the LCR since early 2004. Our review thus far has identified several issues associated with the collection and management of monitoring samples and calculation of the 90th percentile for compliance. The memo reiterates requirements of the regulation and clarifies several areas where there has been confusion.

The Agency is continuing to carry out its national review of implementation, which is aimed at determining whether changes are needed to existing guidance or regulations. The national review includes evaluation of the data USEPA collects under the LCR and an analysis of how states are implementing the rule. As part of the review, national expert workshops were held on monitoring, lead service line replacement, public education, and simultaneous compliance. USEPA is also working with state and local authorities related to monitoring for lead in schools and daycare facilities.

To assure corrosion control treatment technique requirements are effective in protecting public health, the rule also established an AL of 15 ppb for lead in drinking water. Systems are required to monitor a specific number of customer taps, according to the size of the system. Results of monitoring are used to determine the concentration at the 90th percentile (e.g., if 100 samples collected, the concentration at the 90th highest sample). If the 90th percentile exceeds 15 ppb, the system must undertake a number of additional actions to control corrosion and to inform the public about steps they should take to protect their health.

USEPA's review of state programs and press reports have identified inconsistencies in how utilities and states are carrying out the regulation. Although USEPA is carrying out an extensive review to determine if changes to guidance or regulations are needed, it was clear that there was confusion about the existing requirements. A decision was made to release a memo at this time to remind states and utilities of the requirements and to clarify several areas in which there appears to be some confusion with respect to those requirements.

The memorandum answers the following questions, making reference to regulatory citations. The answers below are simplistic summaries of the full responses to the questions. Interested readers are recommended to refer to the seven-page memorandum for complete answers.

*This fact sheet, EPA 810-F-04-001, can be found at: www.epa.gov/ogwdwooo/lcrmr-guide_fs.html.

What samples are used to calculate the 90th percentile?

The memo indicates that results from all samples that are part of a system's targeted sampling pool (sites with the greatest potential for lead leaching) must be used for the calculation of the 90th percentile.

What should utilities do with sample results from customer-request sampling programs?

The memo indicates that samples collected under these programs should not be used to calculate the 90th percentile, except in cases where the system is reasonably able to determine that the site selection criteria for compliance sampling are satisfied.

What should states do with samples taken outside of the sampling compliance period?

The memo indicates that only those samples collected during the compliance monitoring period may be included in the 90th percentile calculation. However, samples collected outside the sampling compliance period must still be provided to the state.

What should states do to calculate compliance if the minimum number of samples is not collected?

States must calculate the 90th percentile even if the minimum number of samples is not collected. A system that fails to collect the minimum required number of samples incurs a monitoring and reporting violation and is thus required to conduct Public Notification.

What is a proper sample?

The memo reiterates that the rule defines a proper sample as a first-draw sample, 1 liter in volume, that is taken after water has been standing in plumbing for at least 6 hours and from an interior tap typically used for consumption—cold-water kitchen or bathroom sink tap in residences. There is no outer limit on standing time.

How can utilities avoid problems with sample collection?

The memo recommends steps utilities can take to avoid analysis of improper samples.

On what grounds may a sample be invalidated?

The memo reiterates the criteria that allow a sample result to be invalidated and makes the point that sample results cannot be invalidated based on homeowner sampling error.

Background on the Lead and Copper Rule

The LCR has four main functions, to: (1) require water suppliers to optimize their treatment system to control corrosion in customers' plumbing; (2) determine tap water levels of lead and copper for customers who have lead service lines or lead-based solder in their plumbing system; (3) rule out the source water as a source of significant lead levels; and (4) if action levels are exceeded, require the suppliers to educate their customers about lead and suggest actions they can take to reduce their exposure to lead through public notices and public education programs. If after installing and optimizing corrosion control treatment, a water system continues to fail to meet the lead action level, it must begin replacing the lead service lines under its ownership. Large systems serving more than 50,000 people were required to conduct studies of corrosion control and to install the state-approved optimal corrosion control treatment by Jan. 1, 1997. Small- and medium-sized systems are required to optimize corrosion control when monitoring at the consumer taps shows action is necessary.

STAGE 1 DISINFECTANTS AND DISINFECTION BY-PRODUCTS RULE

Overview	
Title	Stage 1 Disinfectants and Disinfection By-products Rule (Stage 1 D/DBPR): 63 FR 69390–69476, Dec. 16, 1998, Vol. 63, No. 241 Revisions to the Interim Enhanced Surface Water Treatment Rule (IESWTR), the Stage 1 Disinfectants and Disinfection By-products Rule (Stage 1 D/DBPR), and Revisions to State Primacy Requirements to Implement the Safe Drinking Water Act (SDWA) Amendments: 66 FR 3770, Jan. 16, 2001, Vol. 66, No. 29
Purpose	Improve public health protection by reducing exposure to DBPs. Some disinfectants and DBPs have been shown to cause cancer and reproductive effects in lab animals and suggested bladder cancer and reproductive effects in humans.
General Description	The Stage 1 D/DBPR is the first of a staged set of rules that will reduce the allowable levels of DBPs in drinking water. The new rule establishes seven new standards and a treatment technique of enhanced coagulation or enhanced softening to further reduce DBP exposure. The rule is designed to limit capital investments and avoid major shifts in disinfection technologies until additional information is available on the occurrence and health effects of DBPs.
Utilities Covered	The Stage 1 D/DBPR applies to all sizes of community water systems and nontransient, noncommunity water systems that add a disinfectant to the drinking water during any part of the treatment process and transient, noncommunity water systems that use chlorine dioxide.

Public Health Benefits	
Implementation of the Stage 1 D/DBPR will result in…	• As many as 140 million people receiving increased protection from DBPs. • 24 percent average reduction nationally in trihalomethane levels. • Reduction in exposure to the major DBPs from use of ozone (DBP = bromate) and chlorine dioxide (DBP = chlorite).
Estimated impacts of the Stage 1 D/DBPR include…	• National capital costs: $2.3 billion. • National total annualized costs to utilities: $684 million. • 95 percent of households will incur an increase of less than $1 per month. • 4 percent of households will incur an increase of $1–10 per month. • <1 percent of households will incur an increase of $10–33 per month.

Critical Deadlines and Requirements	
For Drinking Water Systems	
Jan. 1, 2002	Surface water systems and groundwater systems under the direct influence of surface water serving ≥10,000 people must comply with the Stage 1 D/DBPR requirements.
Jan. 1, 2004	Surface water systems and groundwater systems under the direct influence of surface water serving <10,000 and all groundwater systems must comply with the Stage 1 D/DBPR requirements.
For States	
Dec. 16, 2000	States submit Stage 1 D/DBPR primacy revision applications to USEPA (triggers interim primacy).
Dec. 16, 2002	Primacy extension deadline—all states with an extension must submit primacy revision applications to USEPA.

Regulated Contaminants/Disinfectants

Regulated Contaminant	MCL, mg/L	MCLG, mg/L	Regulated Disinfectant	MRDL,* mg/L	MRDLG,* mg/L
Total Trihalomethanes (TTHM)	0.080		Chlorine	4.0 as Cl_2	4
Chloroform		—			
Bromodichloromethane		zero			
Dibromochloromethane		0.06			
Bromoform		zero			
Five Haloacetic Acids (HAA5)	0.060		Chloramines	4.0 as Cl_2	4
Monochloroacetic acid		—	Chlorine dioxide	0.8	0.8
Dichloroacetic acid		zero			
Trichloroacetic acid		0.3			
Bromoacetic acid		—			
Dibromoacetic acid		—			
Bromate (plants that use ozone)	0.010	zero			
Chlorite (plants that use chlorine dioxide)	1.0	0.8			

*Stage 1 D/DBPR includes maximum residual disinfectant levels (MRDLs) and maximum residual disinfectant level goals (MRDLGs), which are similar to MCLs and MCLGs, but for disinfectants.

Treatment Technique

Enhanced coagulation/enhanced softening to improve removal of DBP precursors (see Step 1 TOC table below) for systems using conventional filtration treatment.

Step 1 TOC Table—Required Percentage Removal of TOC

Source Water TOC, *mg/L*	Source Water Alkalinity, *mg/L as* $CaCO_3$		
	0–60	60–120	>120
>2.0 to 4.0	35.0%	25.0%	15.0%
>4.0 to 8.0	45.0%	35.0%	35.0%
>8.0	50.0%	40.0%	30.0%

NOTE: Systems meeting at least one of the alternative compliance criteria in the rule are not required to meet the removals in this table. Systems practicing softening must meet the TOC removal requirements in the last column.

Routine Monitoring Requirements

	Coverage	Monitoring Frequency	Compliance
TTHM/HAA5	Surface and groundwater under the direct influence of surface water serving ≥10,000	4/plant/quarter	Running annual average
	Surface and groundwater under the direct influence of surface water serving 500–9,999	1/plant/quarter	Running annual average
	Surface and groundwater under the direct influence of surface water serving <500	1/plant/year in month of warmest water temperature*	Running annual average of increased monitoring
	Groundwater serving ≥10,000	1/plant/quarter	Running annual average
	Groundwater serving <10,000	1/plant/year in month of warmest water temperature*	Running annual average of increased monitoring
Bromate	Ozone plants	Monthly	Running annual average
Chlorite	Chlorine dioxide plants	Daily at entrance to distribution system; monthly in distribution system	Daily/follow-up monitoring
Chlorine dioxide	Chlorine dioxide plants	Daily at entrance to distribution system	Daily/follow-up monitoring
Chlorine/ chloramines	All systems	Same location and frequency as TCR sampling	Running annual average
DBP precursors	Conventional filtration	Monthly for total organic carbon and alkalinity	Running annual average

*System must increase monitoring to one sample per plant per quarter if an MCL is exceeded.

STAGE 2 DISINFECTANTS AND DISINFECTION BY-PRODUCTS RULE

This section provides essential information on the Stage 2 Disinfectants and Disinfection By-products Rule (Stage 2 D/DBPR) that was promulgated on Jan. 4, 2006, particularly as it applies to all community and nontransient-noncommunity water systems that either add a primary or residual disinfectant other than UV light or deliver such treated water (groundwater or surface water) and serve 50,000 or more people. These systems have the earliest compliance deadlines, which are detailed here.

Implementation of the rule, which complements the Stage 1 D/DBPR, begins with the implementation of an initial distribution system evaluation (IDSE) to identify monitoring locations for eventual compliance with current standards for total trihalomethanes (TTHM) and five haloacetic acids (HAA5) on a locational running annual average (LRAA).

As detailed here, water systems have two ways to obtain IDSE waivers (an automatic waiver for very small systems and a "40/30" certification) and two ways to complete an IDSE (standard monitoring or a system-specific study). Also, nontransient-noncommunity systems serving fewer than 10,000 people do not need to complete any of the four IDSE options.

Complying with the standards at each of several locations determined by the IDSE to have typically elevated TTHM and HAA5 levels on an LRAA basis instead of the Stage 1 D/DBPR's systemwide RAA approach is intended to ensure that all customers receive the same quality water. Under the current approach, the potential exists for some customers to receive water that exceeds the current DBP standards even when the system as a whole complies with the systemwide standard. Reducing this potential is one of the Stage 2 D/DBPR's major objectives.

For purposes of implementing the rule on a staggered schedule, USEPA has placed utilities in one of four compliance "schedules" according to system size.

Also for this rule, USEPA has adopted a standardized national approach for regulating wholesale and consecutive systems that are part of an interconnected distribution system. As defined by the rule,

- A combined distribution system is the interconnected distribution system consisting of the distribution systems of wholesale systems and of the consecutive systems that receive finished water.
- A wholesale system is a public water system that treats source water as necessary to produce finished water and then delivers some or all of that finished water to another public water system. Delivery may be through a direct connection or through the distribution system of one or more consecutive systems.
- A consecutive system is a public water system that receives some or all of its finished water from one or more wholesale systems. Delivery may be through a direct connection or through the distribution system of one or more consecutive systems.

The Stage 2 D/DBPR also requires both IDSE and compliance monitoring locations and frequency to be based on population served and source water type rather than on the number of treatment plants, as under the Stage 1 D/DBPR.

The schedule groups are defined as follows:

- *Schedule 1* systems include affected systems serving 100,000 or more people or those that are part of a combined distribution system in which the largest serves 100,000 or more people.

- *Schedule 2* systems include affected systems serving 50,000 to 99,999 people or those that are part of a combined distribution system in which the largest system serves 50,000 to 99,999 people.
- *Schedule 3* systems include affected systems serving 10,000 to 49,999 people or those that are part of a combined distribution system in which the largest system serves 10,000 to 49,999 people.
- *Schedule 4* systems include affected systems that serve fewer than 10,000 people and are not connected to a larger system.

All but very small systems that are automatically waived from completing an IDSE (and nontransient-noncommunity systems serving fewer than 10,000 people) must submit either a 40/30 certification or a plan for conducting an IDSE by completing standard monitoring or a system-specific study according to the compliance deadlines for each schedule group. Consecutive systems must coordinate IDSE planning and implementation with other systems in the combined distribution system but must conduct their own IDSE based on individual system size.

USEPA has developed an online compliance assistance tool to help utilities select and implement the appropriate IDSE option from among the four offered. The IDSE tool comprises two modules. One, dubbed "The Wizard," prompts users to answer a series of questions intended to identify the appropriate IDSE option. The other provides users with custom electronic forms (by system size and type) to prepare an IDSE plan and/or report.

The IDSE tool is a component of an electronic Data Collection and Tracking System USEPA has developed to collect and manage information generated by utilities under both the Stage 2 D/DBPR and companion Long-Term 2 Enhanced Surface Water Treatment Rule (LT2ESWTR), which was promulgated on Jan. 5, 2006.

Systems that complete standard monitoring or a system-specific study also must submit a final IDSE report according to their respective schedule group deadlines, and all systems must also meet deadlines for developing and implementing a routine compliance monitoring plan. The rule provides a step-by-step process for selecting compliance monitoring locations.

The Stage 2 D/DBPR also requires each system to determine if they have exceeded an operational evaluation level, which is identified using their compliance monitoring results. The operational evaluation level provides an early warning of possible future violations of DBP standards, which allows the system to take proactive steps to remain in compliance. A system that exceeds an operational evaluation level is required to review its operational practices and submit a report to the state identifying actions that may be taken to mitigate future high DBP levels, particularly those that may jeopardize their compliance with the DBP standards.

The key Stage 2 D/DBPR compliance deadlines, by schedule group, are listed in the following table:

Stage 2 D/DBPR key compliance deadlines by system size schedule

Schedule	Submit IDSE Plan*	Complete IDSE or System-Specific Study	Submit IDSE Report	Comply with DBP Standards on LRAA Basis
Schedule 1 systems (serving ≥100,000)	Oct. 1, 2006	Sept. 30, 2008	Jan. 1, 2009	Apr. 1, 2012
Schedule 2 systems (serving 50,000–99,999)	Apr. 1, 2007	Mar. 31, 2009	July 1, 2009	Oct. 1, 2012
Schedule 3 systems (serving 10,000–49,999)	Oct. 1, 2007	Sept. 30, 2009	Jan. 1, 2010	Oct. 1, 2013
Schedule 4 systems (serving <10,000)	Apr. 1, 2008	Mar. 31, 2010	July 1, 2010	Oct. 1, 2013†

*40/30 certification, standard monitoring plan, or system-specific study plan.
†This is the compliance date for systems required to monitor for *Cryptosporidium* under the companion LT2ESWTR. The deadline for systems that are not required to monitor for *Cryptosporidium* is Oct. 1, 2014. Also, states may grant up to an additional 2 years to systems making capital improvements.

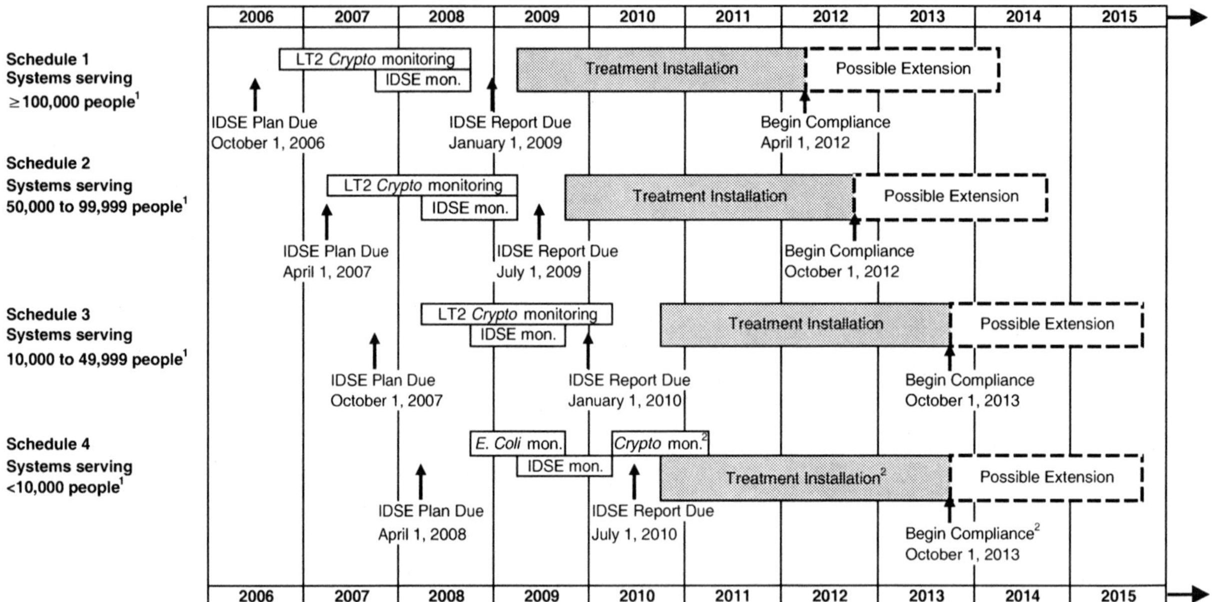

Figure S2DBPR-1 Stage 2 D/DBPR and LT2ESWTR compliance schedule

Also under the Stage 2 DBPR, consecutive systems that deliver water treated with a disinfectant other than UV must comply with chlorine and chloramines monitoring requirements and maximum residual disinfectant levels beginning Jan. 1, 2009 (or earlier if required by the state).

Finally, USEPA has prepared many compliance assistance resources, including guidance documents on IDSE requirements, operational evaluations, consecutive systems, and simultaneous compliance with the LT2ESWTR and other regulations. These compliance assistance tools are online at www.epa.gov/ogwdw/disinfection/stage2/index.html.

Figure S2DBPR-1 provides a comprehensive compliance schedule for both the LT2ESWTR and the Stage 2 D/DBPR.

Overview	
Title	Stage 2 Disinfectants and Disinfection By-products Rule (Stage 2 D/DBPR); 71 FR 388, Jan. 4, 2006, Vol. 71, No. 2
Purpose	To increase public health protection by reducing the potential risk of adverse health effects associated with disinfection by-products (DBPs) throughout the distribution system. Builds on the Stage 1 Disinfectants and Disinfection By-products Rule (Stage 1 D/DBPR) by focusing on monitoring for and reducing concentrations of two classes of DBPs—TTHM and HAA5—in drinking water.
General Description	Stage 2 D/DBPR requires systems to complete an initial distribution system evaluation (IDSE) to characterize DBP levels in their distribution systems and identify locations to monitor DBPs for Stage 2 D/DBPR compliance. The Stage 2 D/DBPR bases TTHM and HAA5 compliance on a locational running annual average (LRAA) calculated at each monitoring location.
Utilities Covered	• All community water systems (CWSs) and nontransient noncommunity water systems (NTNCWSs) that either add a primary or residual disinfectant other than UV light, or deliver water that has been treated with a primary or residual disinfectant other than UV light. • Schedule 1 includes CWSs and NTNCWSs serving 100,000 or more people OR CWSs and NTNCWSs that are part of a combined distribution system in which the largest system serves 100,000 or more people.

Stage 2 D/DBPR Regulated Contaminants

Regulated Contaminants	MCLG (mg/L)	MCL (mg/L)
Total Trihalomethanes (TTHM)		0.080 LRAA
Chloroform Bromodichloromethane Dibromochloromethane Bromoform	0.07 zero 0.06 zero	
Five Haloacetic Acids (HAA5)		0.060 LRAA
Monochloroacetic acid Dichloroacetic acid Trichloroacetic acid Bromoacetic acid Dibromoacetic acid	0.07 zero 0.02 — —	

IDSE Requirements*	
IDSE Option	**Description**
Standard Monitoring	Standard monitoring is one year of increased monitoring for TTHM and HAA5 in addition to the data being collected under Stage 1 D/DBPR. These data will be used with Stage 1 D/DBPR data to select Stage 2 D/DBPR monitoring locations for DBP compliance monitoring. Any system may conduct standard monitoring to meet the IDSE requirements of the Stage 2 D/DBPR.
System-Specific Study (SSS)	Systems that have extensive DBP data (including Stage 1 D/DBPR compliance data) or technical expertise to prepare a hydraulic model may choose to conduct a system specific study to select Stage 2 D/DBPR compliance monitoring locations.
40/30 Certification†	The term "40/30" refers to a system that during a specific time period has all individual Stage 1 D/DBPR compliance samples less than or equal to 0.040 mg/L for TTHM and 0.030 mg/L for HAA5 and has no monitoring violations during the same time period. These systems have no IDSE monitoring requirements, but will still need to conduct Stage 2 D/DBPR compliance monitoring.
Very Small System (VSS) Waiver†	Systems that serve fewer than 500 people and have DBP data can qualify for a VSS waiver and would not be required to conduct IDSE monitoring. These systems have no IDSE monitoring requirements, but will still need to conduct Stage 2 D/DBPR compliance monitoring.
USEPA has developed several tools to assist systems with complying with the Stage 2 D/DBPR IDSE requirements. These materials can be downloaded at www.epa.gov/safewater/disinfection/stage2.	
*NTNCWS serving <10,000 people do not need to complete any of the IDSE options. †Unless notified by USEPA or the state that they must complete standard monitoring or system-specific study.	

Compliance With Stage 2 D/DBPR MCLs (Routine Monitoring)			
Source Water Type	**Population Size Category**	**Monitoring Frequency***	**Total Distribution System Monitoring Locations per Monitoring Period†**
Subpart H	<500	per year	2
	500–3,300	per quarter	2
	3,301–9,999	per quarter	2
	10,000–49,999		4
	50,000–249,999		8
	250,000–999,999		12
	1,000,000–4,999,999		16
	≥5,000,000		20

Table continued next page.

Compliance With Stage 2 D/DBPR MCLs (Routine Monitoring) (continued)

Source Water Type	Population Size Category	Monitoring Frequency*	Total Distribution System Monitoring Locations per Monitoring Period†
Ground Water	<500	per year	2
	500–9,999		2
	10,000–99,999	per quarter	4
	100,000–499,999		6
	≥500,000		8

Operational Evaluation

Systems must begin complying with rule requirements to determine compliance with the operational evaluation levels for TTHMs and HAA5s.

*All systems must monitor during month of highest DBP concentrations.

†Systems on quarterly monitoring must take dual sample sets every 90 days at each monitoring location, except for subpart H systems serving 500–3,300. Systems on annual monitoring and subpart H systems serving 500–3,300 are required to take individual TTHM and HAA5 samples (instead of a dual sample set) at the locations with the highest TTHM and HAA5 concentrations, respectively. If monitoring annually, only one location with a dual sample set per monitoring period is needed if highest TTHM and HAA5 concentrations occur at the same location, and month.

Critical Deadlines and Requirements

For Drinking Water Systems

Schedule 1 Deadlines	Schedule 2 Deadlines	Requirements
Jan. 4, 2006	Jan. 4, 2006	Systems serving fewer than 500 people that have TTHM and HAA5 compliance data qualify for a VSS waiver from conducting an IDSE, unless informed otherwise by USEPA or state primacy agency.
Oct. 1, 2006	Apr. 1, 2007	Systems that do not receive a VSS waiver must submit to the USEPA or state primacy agency either a: • Standard monitoring plan, • System-specific study plan, or • 40/30 certification.
Oct. 1, 2007	Apr. 1, 2008	Systems conducting standard monitoring or SSS begin collecting samples in accordance with their approved plan.
Sept. 30, 2008	Mar. 31, 2009	No later than this date, systems conducting standard monitoring or a SSS complete their monitoring or study.
Jan. 1, 2009	July 1, 2009	No later than this date, systems conducting standard monitoring or a SSS must submit their IDSE report.
Apr. 1, 2009	Apr. 1, 2009	Consecutive systems must begin monitoring for chlorine or chloramines as specified under the Stage 1 D/DBPR.

Table continued next page.

Critical Deadlines and Requirements (continued)		
For Drinking Water Systems		
Schedule 1 Deadlines	**Schedule 2 Deadlines**	**Requirements**
Apr. 1, 2012	Oct. 1, 2012	No later than this date, systems must: • Complete their Stage 2 D/DBPR compliance monitoring plan (systems serving more than 3,300 people must submit their monitoring plan to the state.)* • Begin complying with monitoring requirements of the Stage 2 D/DBPR.†
January 2013	July 2013	Systems must begin complying with rule requirements to determine compliance with the operational evaluation levels for TTHMs and HAA5s.
For States		
Schedule 1 Deadlines	**Schedule 2 Deadlines**	**Requirements**
January–June 2006	January–June 2006	States are encouraged to inform systems serving fewer than 500 people and do not qualify for a VSS waiver from the IDSE requirements should begin complying with standard monitoring requirements.
Sept. 30, 2007	Mar. 31, 2008	States must approve the system's standard monitoring plan, 40/30 certification, or system-specific study plan, or notify the system that the state has not completed its review.
Oct. 4, 2007	Oct. 4, 2007	States are encouraged to submit final primacy applications or extension requests to USEPA.
Jan. 4, 2008	Jan. 4, 2008	Final primacy applications must be submitted to USEPA, unless granted an extension.
Mar. 31, 2009	Sept. 30, 2009	States must approve the system's IDSE report or notify the system that the state has not completed its review of the IDSE report.
Jan. 4, 2010	Jan. 4, 2010	Final primacy revision applications from states with approved 2-year extensions agreements must be submitted to USEPA.

*A monitoring plan is not required if the IDSE report includes all information required in the monitoring plan.
†States may allow up to an additional 24 months for compliance with MCLs for systems requiring capital improvements.

RADIONUCLIDES RULE

Overview	
Title	Radionuclides Rule: 66 FR 76708; Dec. 7, 2000, Vol. 65, No. 236
Purpose	Reducing the exposure to radionuclides in drinking water will reduce the risk of cancer. This rule will also improve public health protection by reducing exposure to all radionuclides.
General Description	The rule retains the existing MCLs for combined radium-226 and radium-228, gross alpha particle radioactivity, and beta particle and photon activity. The rule regulates uranium for the first time.
Utilities Covered	CWSs, all size categories.

Public Health Benefits	
Implementation of the Radionuclides Rule will result in…	Reduced uranium exposure for 620,000 persons, protection from toxic kidney effects of uranium, and a reduced risk of cancer.
Estimated impacts of the Radionuclides Rule include…	Annual compliance costs of $81 million. Only 795 systems will have to install treatment.

Regulated Contaminants

Regulated Radionuclide	MCL	MCLG
Beta/photon emitters*	4 mrem/year	0
Gross alpha particle	15 pCi/L	0
Combined radium-226/228	5 pCi/L	0
Uranium	30 pg/L	0

NOTE: MCLG—maximum contaminant level goal.
*A total of 168 individual beta particle and photon emitters may be used to calculate compliance with the MCL.

Critical Deadlines and Requirements	
For Drinking Water Systems	
June 2000–Dec. 8, 2003	When allowed by the state, data collected between these dates may be eligible for use as grandfathered data (excluding beta particle and photon emitters).
Dec. 8, 2003	Systems begin initial monitoring under state-specified monitoring plan unless the state permits use of grandfathered data.
Dec. 31, 2007	All systems must complete initial monitoring.
For States	
December 2000–December 2003	States work with systems to establish monitoring schedules.
Dec. 8, 2000	States should begin to update vulnerability assessments for beta photon and particle emitters and notify systems of monitoring requirements.

Table continued next page.

Critical Deadlines and Requirements (continued)	
For States (continued)	
Spring 2001	USEPA meets and works with states to explain new rules and requirements and to initiate adoption and implementation activities.
Dec. 8, 2002	State submits primacy revision application to USEPA (USEPA approves within 90 days).

Monitoring Requirements	
Gross Alpha, Combined Radium —226/228, and Uranium*	Beta Particle and Photon Radioactivity*
Initial Monitoring	
Four consecutive quarters of monitoring	No monitoring required for most CWSs. Vulnerable CWSs† must sample for: • Gross beta: quarterly samples • Tritium and strontium-90: annual samples.
Reduced Monitoring	
If the average of the initial monitoring results for each contaminant is below the detection limit: one sample every 9 years. If the average of the initial monitoring results for each contaminant is greater than or equal to the detection limit but less than or equal to one-half the MCL: one sample every 6 years. If the average of the initial monitoring results for each contaminant is greater than one-half the MCL but less than or equal to the MCL: one sample every 3 years.	If the running annual average of the gross beta particle activity minus the naturally occurring potassium-40 activity is less than or equal to 50 pCi/L: one sample every 3 years.
Increased Monitoring	
A system with an entry point result above the MCL must return to quarterly sampling until four consecutive quarterly samples are below the MCL.	If gross beta particle activity minus the naturally occurring potassium-40 activity exceeds 50 pCi/L, the system must: • Speciate as required by the state • Sample at the initial monitoring frequency.

*All samples must be collected at each entry point to the distribution system.
†The rule also contains requirements for CWSs using waters contaminated by effluents from nuclear facilities.

Grandfathering of Data

When allowed by the state, data collected between June 2000 and Dec. 8, 2003, may be used to satisfy the initial monitoring requirements if samples have been collected from:

- Each EPTDS.
- The distribution system, provided the system has a single EPTDS.
- The distribution system, provided the state makes a written justification explaining why the sample is representative of all EPTDS.

PROPOSED RADON IN DRINKING WATER RULE

In November 1999 USEPA proposed new regulations (64 FR 5924b) to protect people from exposure to radon. The proposed regulations will provide states flexibility in how to limit the public's exposure to radon by focusing their efforts on the greatest public health risks from radon—those in indoor air—while also reducing the highest risks from radon in drinking water. The framework for this proposal is set out in the Safe Drinking Water Act as amended in 1996, which provides for a multimedia approach to address the public health risks from radon in drinking water and radon in indoor air from soil. The Safe Drinking Water Act directs the USEPA to propose and finalize a maximum contaminant level (MCL) for radon in drinking water, but also to make available an alternative approach: a higher alternative maximum contaminant level (AMCL) accompanied by a multimedia mitigation (MMM) program to address radon risks in indoor air. This framework reflects the unique characteristics of radon: in most cases, radon released to indoor air from soil under homes and buildings is the main source of exposure and radon released from tap water is a much smaller source of radon in indoor air. It is more cost-effective to reduce risk from radon exposure from indoor air than from drinking water. USEPA strongly encourages states to take full advantage of the flexibility and risk reduction opportunities in the MMM program.

What Are the Public Health Concerns?

Radon is a naturally occurring radioactive gas that emits ionizing radiation. National and international scientific organizations have concluded that radon causes lung cancer in humans. Most of the radon in indoor air comes from the breakdown of uranium in soil beneath homes. Breathing radon from the indoor air in homes is the primary public health risk from radon, contributing to about 20,000 lung cancer deaths each year in the United States, according to a 1999 landmark report by the National Academy of Sciences (NAS) on radon in indoor air. The US Surgeon General has warned that radon in indoor air is the second leading cause of lung cancer. USEPA and the US Surgeon General recommend testing all homes and apartments located below the third floor for radon in indoor air. If you smoke and your home has high indoor radon levels, your risk of lung cancer is especially high.

Radon from tap water is a smaller source of radon in indoor air. Only about 1–2 percent of radon in indoor air comes from drinking water. However, breathing radon released to air from household water use increases the risk of lung cancer over the course of your lifetime. Ingestion of drinking water containing radon also presents a risk of internal organ cancers, primarily stomach cancer. This risk is smaller than the risk of developing lung cancer from radon released to air from tap water. Based on a second 1999 NAS report on radon in drinking water, USEPA estimates that radon in drinking water causes about 168 cancer deaths per year, 89 percent from lung cancer caused by breathing in radon released from water, and 11 percent from stomach cancer caused by drinking radon-containing water.

Who Must Comply with the Proposed Rule?

The proposed radon in drinking water rule applies to all community water systems (CWSs) that use groundwater or mixed ground and surface water (e.g., systems serving homes, apartments, and trailer parks). The proposed rule would not apply to CWSs that use solely surface water, nor to nontransient noncommunity public water supplies and transient public water supplies (e.g., systems serving schools, office buildings, campgrounds, restaurants, and highway reststops).

What Does the Rule Propose to Require?

The rule proposes a maximum contaminant level goal (MCLG), an MCL, an AMCL, and requirements for MMM program plans to address radon in indoor air. The proposal also includes monitoring, reporting, public notification and consumer confidence report requirements, proposed best available technologies and analytical methods.

Maximum Contaminant Level Goal (MCLG), Maximum Contaminant Level (MCL), and Alternative Maximum Contaminant Level (AMCL)

The proposed MCLG for radon in drinking water is zero. This is a nonenforceable goal.

The proposed regulation provides two options for the maximum level of radon that is allowable in community water supplies. The proposed MCL is 300 picoCuries per liter (pCi/L) and the proposed AMCL is 4,000 pCi/L. The drinking water standard that would apply for a system depends on whether or not the state or CWS develops a MMM program. The regulatory expectation of CWSs serving 10,000 persons or less is that they meet the 4,000-pCi/L AMCL and be associated with an approved MMM program plan—either developed by the state or by the CWS.

The enforceable MCL or AMCL would apply under the following circumstances:

Small CWSs: Proposed Regulatory Expectation for Systems That Serve 10,000 or Fewer People

Does State develop MMM program?	Does CWS develop local MMM program?	CWS Complies with:
Yes	Not needed	AMCL: 4000 pCi/L*
>No	Yes**	AMCL: 4000 pCi/L

* Small systems may elect to comply with the MCL of 300 pCi/L
** Small systems may elect to comply with the MCL of 300 pCi/L, instead of developing a local MMM program.

Large CWSs: Proposed Compliance Options for Systems That Serve More Than 10,000 People

Does State develop MMM program?	Does CWS develop local MMM program?	CWS Complies with:
Yes	Not needed	AMCL: 4000 pCi/L*
No	Yes	MCL: 300 pCi/L
No	No	AMCL: 4000 pCi/L

* Large systems may elect to comply with the MCL of 300 pCi/L

Monitoring Requirements

CWSs must monitor for radon in drinking water according to the requirements described in the table below and report their results to the State. If the State determines that the radon level in a CWS is below 300 pCi/L, the system only needs to continue meeting monitoring requirements and is not covered by the requirements regarding MMM programs.

Type	Frequency	Condition
Initial	Four consecutive quarters of monitoring for one year.	At each entry point to the distribution system that is representative of each well after treatment and/or storage
Routine	One sample per year	If running average from four consecutive quarterly samples is less than MCL/AMCL, and at the discretion of State.
Reduced	One sample every three years	If average from four consecutive quarterly samples is less than ½ the MCL/AMCL, no samples exceed the MCL/AMCL, and State determines the system is "reliably and consistently below MCL/AMCL."
Increased	Four consecutive quarters of monitoring	If the MCL/AMCL for radon is exceeded in a single sample, when monitoring annually. Can return to one sample per year if meet routine monitoring conditions, listed above.

Promulgation of the final Radon Rule has been repeatedly delayed by USEPA, which in 2005 indicated that the rule will be promulgated in December 2006.

Summaries of Other Regulations

UNREGULATED CONTAMINANT MONITORING RULE

To help identify chemical and microbial drinking water contaminants that may require future regulation, the 1996 amendments to the SDWA required the USEPA to establish regulations for an ongoing unregulated contaminant monitoring program.

USEPA is to use the data generated by a periodically updated regulation to evaluate and prioritize contaminants on the Drinking Water Contaminant Candidate List and to support regulatory determinations and future rulemakings. The Drinking Water Contaminant Candidate List, which is periodically updated, identifies contaminants the agency considers for possible new drinking water standards.

USEPA promulgated the first Unregulated Contaminant Monitoring Rule (UCMR) in 1999 and updated it in 2000 and again in 2001. It identified 36 unregulated contaminants for monitoring and categorized them according to the availability of suitable analytical methods (Table UCMR-1). USEPA defined a suitable method as "one with a proven track record of providing consistent, quality data on the occurrence of the analyte, and whose cost would not prohibit its use on a national scale."

Those with available suitable methods were identified for immediate assessment monitoring by all large water systems and a representative sample of small systems. Contaminants with newly developed methods requiring additional assessment were subject to screening surveys, and those with methods requiring research and development were subject to prescreen testing.

The rule also included requirements for:

- all large PWSs and a representative sample of small PWSs to monitor for List 1 contaminants;
- selected large and small PWSs to monitor for List 2 contaminants;
- affected water systems to submit monitoring data to USEPA and the states for inclusion in the NCOD;
- affected water systems to notify consumers of the availability of the results of UCMR monitoring and include detected contaminants in annual Consumer Confidence Reports; and
- USEPA to pay for the reasonable testing costs for the representative sample of small systems.

Per SDWA requirements, the UCMR provides for states to develop monitoring plans for small systems included in the national representative sample. The UCMR also provides for states to establish partnership agreements with USEPA to accept or modify the initial monitoring plan, determine an alternative vulnerable monitoring time, modify the timing of monitoring to coordinate with compliance monitoring, identify alternative sampling points, notify systems of monitoring responsibilities, provide instructions to small systems, participate in the List 2 survey and List 3 prescreen testing, and provide additional locational information for systems.

Second UCMR Round Coming

USEPA proposed requirements for the second round of UCMR monitoring in August 2005. Under that proposal, which is expected to be adopted in early 2006, a national sampling of water utilities will be required to monitor for 26 high-priority unregulated contaminants, including perchlorate again. The second UCMR (UCMR2) comprises an assessment monitoring component that would have 3,910 public water systems look for 11 of the targeted chemicals using six well-established analytical methods. It also includes a screening survey that would have 1,122 systems monitor for the other 15 chemicals using specialized and less commonly used methods.

Unlike the first UCMR (UCMR1) program, the second round of unregulated contaminant monitoring does not target any microbial contaminants and does not include a prescreening component (UCMR1 prescreening was never completed for lack of available methods).

In addition to targeting many new contaminants for monitoring, the proposed UCMR2 differs from its predecessor by

- expanding the number of systems required to conduct the screening survey;
- revising laboratory-approval and data-reporting requirements to reflect lessons learned during UCMR1 implementation;
- eliminating use of "index" system monitoring;
- limiting assessment monitoring to distribution system entry points; and
- establishing a firm utility-applicability cutoff date of June 30, 2005.

The proposed rule would also clarify the "population served" definition to include customers of any consecutive systems and define a protocol for establishing contaminant-specific minimum reporting levels (MRLs) in lieu of detection limits. Key aspects of UCMR2 that remain identical to UCMR1 include direct implementation by USEPA (with voluntary state participation); USEPA covering the cost of small system sample kits, shipping, and analysis; and the basic statistical framework for assessment monitoring.

Also as before, large-system monitoring must be conducted using designated methods by laboratories approved by USEPA through a process that requires application and proficiency testing. Also, labs must report results (subject to utility review) directly to USEPA via the agency's electronic reporting system.

Assessment Monitoring

As under UCMR1, all 3,110 community and nontransient-noncommunity systems serving more than 10,000 people and a representative sample of 800 such systems serving 10,000 or fewer people will be required to conduct a 12-month period of monitoring for 11 unregulated contaminants during a 3-year window. For UCMR2, the window will be from July 2007 through June 2010, and this time USEPA will require one third of the systems to monitor during each 12-month period to avoid having too many systems do so all in the last 12-month period, as happened under UCMR1.

Of the 800 small systems, 160 will serve 500 or fewer people; 300 will serve 501 to 3,300; and 340 will serve 3,301 to 10,000. Also, there will be at least two participating systems from each state (including at least two Native American systems), and 583 will be groundwater systems and 217 will use surface water.

The 11 targeted contaminants for assessment monitoring include two pesticide chemicals (dimethoate and terbufos sulfone), five flame retardants (BDE-47, -99, -100, and -153 and 245-HBB), three explosives (TNT, RDX, and 1.3-dinitrobenzene), and perchlorate (Table UCMR-2).

USEPA said it again selected perchlorate for monitoring to supplement "substantial" UCMR1 data collected using a method accurate to 4 μg/L with data from improved methods that can achieve an MRL of 0.57 μg/L.

Noting that the per-sample cost of the newer methods will be about double that for the UCMR1 method (for a national total of roughly $4.4 million under UCMR2), the agency said the additional data "would provide a more complete understanding of perchlorate's occurrence in drinking water" as it considers possible regulation.

Screening Survey

As proposed, the UCMR2 would have 1,122 systems monitoring for 15 chemicals during one of two 12-month periods between July 2007 and June 2009. Included in the mix would be all of 322 systems serving more than 100,000 people; a randomly selected group of 320 systems serving 10,001 to 100,000; and 480 systems serving 10,000 or fewer. While the systems serving more than 10,000 would also conduct assessment monitoring, the sample of smaller systems chosen to perform the survey would comprise different utilities than those selected for assessment monitoring. Under UCMR1, only 300 systems (randomly selected from among the assessment monitoring pool) participated in the screening survey, which was intended only for limited purposes. USEPA said the statistically stronger UCMR2 screening survey design, in contrast, is intended "to ensure the data can be used to support regulatory determinations and rule development, if warranted."

Chemicals targeted for screening survey monitoring include nine variants of acetanilide herbicides and six nitrosamines, including the disinfection by-product NDMA (Table UCMR-3). Among the herbicides are three parent compounds (acetochlor, alachlor, and metolachlor) and their respective ethane sulfonic acid (ESA) and oxanilic acid (OA) degradation products, which are generally more widespread than their parent compounds. While alachlor is currently regulated, USEPA said it is proposing monitoring of alachlor concurrent with its two degradates "to determine the degree of correlation between the parent compound and degradate occurrence." USEPA noted that it "may consider adding" triazine parent compounds (atrazine, simazine, and propazine) and their three chlorodegradates (desethylatrazine, desisopropylatrazine, and diamino-chlorotriazine) to the UCMR2 screening survey. The agency said a method for monitoring the six chemicals is expected to be ready within the next year, adding that if the six were added, two chemicals would have to be bumped off the UCMR2 list of 26 target contaminants to prevent the total number from exceeding the legal limit of 30.

State Participation

As with UCMR1, states can voluntarily agree to participate in UCMR2 implementation at varying levels, the basic being review of USEPA-generated monitoring plans that will identify selected systems and monitoring schedules. The agency was to have negotiated formal partnership agreements by the end of 2005, with final agreed-upon monitoring plans to be distributed after UCMR2 is finalized in 2006. For states that opt not to partner, USEPA will directly notify selected systems of their monitoring schedules.

Table UCMR-1 UCMR monitoring list

List 1 Assessment Monitoring of Contaminants With Available Methods	List 2 Screening Surveys of Contaminants With Methods Just Developed	List 3 Prescreen Testing of Contaminants Needing Research on Methods
2,4-dinitrotoluene	1,2-diphenylhydrazine	Lead-210
2,6-dinitrotoluene	2-methyl-phenol	Polonium-210
Acetochlor	2,4-dichlorophenol	Cyanobacteria
DCPA mono-acid degradate	2,4-dinitrophenol	Echoviruses
DCPA di-acid degradate	2,4,6-trichlorophenol	Coxsackieviruses
4,4'-DDE	Diazinon	Helicobacter pylori
EPTC	Disulfoton	Microsporidia
Molinate	Diuron	Caliciviruses
MTBE	Fonofos	Adenoviruses
Nitrobenzene	Linuron	
Perchlorate	Nitrobenzene	
Terbacil	Prometon	
	Terbufos	
	Aeromonas	
	Alachlor ESA	
	RDX	

Table UCMR-2 Contaminants and corresponding analytical methods and minimum reporting levels proposed for assessment monitoring*

Contaminant	USEPA Method	Minimum Reporting Level (μg/L)
Dimethoate	527	0.71
Terbufos sulfone	527	0.44
2,2',4,4'-Tetrabromodiphenyl ether (BDE-47)	527	0.33
2,2',4,4',5-Pentabromodiphenyl ether (BDE-99)	527	0.92
2,2',4,4',5,5'-Hexabromobiphenyl (245-HBB)	527	0.72
2,2',4,4',5,5'-Hexabromobiphenyl ether (BDE-153)	527	0.85
2,2',4,4',6-Pentabromodiphenyl ether (BDE-100)	527	0.52
1,3-Dinitrobenzene	529	0.76
2,4,6-Trinitrotoluene (TNT)	529	0.78
Hexahydro-1,3,5-trinitro-1,3,5-triazine (RDX)†	529	1.20
Perchlorate†‡	314.0 (enhanced), 314.1, 331.0, 322.0	0.57

NOTE: USEPA—US Environmental Protection Agency
*Monitoring is limited to entry points to the distribution system (groundwater systems can propose to sample at representative entry points instead of every entry point).
†Listed on second Contaminant Candidate List for regulatory consideration
‡All perchlorate samples must be collected using the sterile technique required in USEPA methods 314.1, 331.0, or 332.0.

Table UCMR-3 Contaminants and corresponding analytical methods and minimum reporting levels proposed for screening survey

Contaminant	USEPA Method	Minimum Reporting Level (μg/L)
Acetochlor*	525.2	2.0
Acetochlor ESA	535	1.4
Acetochlor OA	535	1.5
Alachlor†	525.2	1.6
Alachlor ESA*	535	1.0
Alachlor OA	535	1.6
Metolachlor*	525.2	1.0
Metolachlor ESA	535	1.1
Metolachlor OA	535	1.5
N-nitrosodiethylamine‡	521	0.0046
N-nitrosodimethylamine‡	521	0.0024
N-nitrosodi-*n*-butylamine‡	521	0.0035
N-nitrosodi-*n*-propylamine‡	521	0.0072
N-nitroso-methylethylamine‡	521	0.0034
N-nitroso-pyrrolidine‡	521	0.0022

NOTE: ESA—ethane sulfonic acid, OA—oxanilic acid, USEPA—US Environmental Protection Agency

*Listed on the second Contaminant Candidate List for regulatory consideration (in addition to alachor ESA, the list includes "other acetanilide pesticide degradation products" that are to be determined).

†Although already regulated, alachlor is included to improve understanding of its correlative occurrence with alachor ESA and alachor OA.

‡Samples must be collected at entry points to the distribution system and at distribution system maximum-residence-time locations.

VARIANCES AND EXEMPTIONS

Overview

Title	Variances and Exemptions Rule: 63 FR 43834–43851, Aug. 14, 1998	
	General and Small System Variances	Exemptions
Purpose	Variances allow eligible systems to provide drinking water that does not comply with a NPDWR on the condition that the system installs a certain technology and the quality of the drinking water is still protective of public health.	Exemptions allow eligible systems additional time to build capacity in order to achieve and maintain regulatory compliance with newly promulgated NPDWRs, while continuing to provide acceptable levels of public health protection.
General	There are two types of variances: 1. General variances are intended for systems that are not able to comply with a NPDWR due to their source water quality. 2. Small system variances are intended for systems serving 3,300 persons or fewer that cannot afford to comply with a NPDWR (but may be allowed for systems serving up to 10,000 persons).	Exemptions do not release a water system from complying with NPDWRs; rather, they allow water systems additional time to comply with NPDWRs.
Compliance Date	• *General variances* require compliance as expeditiously as practicable and in accordance with a compliance schedule determined by the state. • *Small system variances* require compliance within 3 years (with a possible 2-year extension period).	Systems must achieve compliance as expeditiously as practicable and in accordance with the schedule determined by the state. In addition: • Initial exemptions cannot exceed 3 years. • Systems serving <3,301 persons may be eligible for one or more additional 2-year extension periods (not to exceed 6 years).
Contaminants Excluded	• General variances may generally not be granted for the MCL for total coliforms or any of the TT requirements of Subpart H of 40 CFR 141. • Small system variances may not be granted for NPDWRs promulgated prior to 1986 or MCLs, indicators, and TTs for microbial contaminants.	• Exemptions from the MCL for total coliforms may generally not be granted.

Utilities Covered

All Public Water Systems	Exclusions: • Systems that have received a small system variance are not eligible for an exemption. • Small system variances may not be granted for NPDWRs that do not list a SSVT. • Systems that have received an exemption are generally not eligible for a variance.

Definitions	
State	For purposes of this document, "state" is used to refer to the primacy agency.
BAT	The BAT, TT, or other means identified by USEPA for use in complying with a NPDWR.
SSVT	A treatment technology identified by USEPA specifically for use by a small public water system that will achieve the maximum reduction or inactivation efficiency that is affordable considering the size of the system and the quality of its source water, while adequately protecting public health.
SSCT	A treatment technology that is affordable by small systems and allows systems to achieve compliance with the requirements of a NPDWR.

Rule-Related Activites and Responsibilities		
	Systems	**States**
General and Small System Variances	• May apply for, if eligible and unable to meet the NPDWR. • Work with the state to hold a public hearing on the proposed variance. • Meet all compliance criteria, including schedule set by the state, once the variance is approved. • Must provide public notice within 1 year after the system begins operating under the variance.	• Review the system's application to determine whether the system meets all eligibility criteria. • Before issuing a variance, determine a schedule for compliance and implementation. • Work with the system to hold a public hearing on the variance and notify USEPA of all variances.
	Systems	**States**
Additional Activities for Small System Variances	• May apply for only if USEPA has identified an SSVT for the rule. • Work with the state to provide notice of the proposed variance to all persons served by the system.	• Determine whether the system is financially and technically able to install and operate an USEPA-approved SSVT. • Work with the system to provide notice of the proposed variance to all persons served by the system and USEPA. • Review all small system variances every 5 years.
Exemptions	• May apply for, if eligible and unable to meet the NPDWR. • Work with the state to hold a public hearing on the proposed exemption. • Upon approval, must meet all compliance criteria and comply with the NPDWR within 3 years. (NOTE: Systems serving <3,301 persons may be eligible for an extension.) • Systems must provide public notice within 1 year after the system begins operating under the exemption.	• Review the system's application to determine whether the system meets all eligibility criteria. • Before issuing an exemption, determine a schedule for compliance and implementation. • Work with the system to hold a public hearing on the exemption and notify USEPA of all exemptions.

General Variances	
Eligibility Requirements	
No Alternative Water Source	Using raw water sources that are reasonably available, the system is unable to meet MCLs [SDWA §1415(a)(1)(A) and 40 CFR 142.40(a)(1)].
Does Not Pose an URTH	The state must determine that the granting of the variance will not pose an URTH [SDWA §1415(a)(1)(A) and 40 CFR 142.40(a)(2)].
Compliance Requirements	
Compliance Date	Systems must comply with the NPDWR as soon as practicable and in accordance with a compliance schedule determined by the state [SDWA §1415(a)(1)(A) and 40 CFR 142.41(c)(4)].
Technology Improvements	The system must install and operate the BAT, TT, or other means found available by USEPA as expeditiously as possible [SDWA §1415(a)(1)(A) and 40 CFR 142.42(c)].
Public Hearing	Before a variance may take effect, the state must provide notice and opportunity for a public hearing on the variance and schedule [SDWA §1415(a)(1)(A) and 40 CFR 142.44].
Public Notification	Systems must provide public notice within 1 year after the system begins operating under a variance and repeat the notice annually for the duration of the variance [40 CFR 141.204(b)(1)].

Small System Variances	
Eligibility Requirements	
System Size	Generally available for systems serving <3,301 persons and, with the approval of USEPA, systems serving >3,300 persons but <10,000 persons [SDWA §1415(e)(1)(A)&(B) and 40 CFR 142.303(a)&(b)].
SSVT	Systems must install, operate, and maintain in accordance with guidance or regulations issued by the USEPA Administrator, a TT or other means that USEPA has identified as a variance technology that is applicable to the size and source water quality conditions of the system [SDWA §1415(e)(2)(A)&(B) and 40 CFR 142.307(b)].
Affordability	In accordance with the affordability criteria established by the state, the system cannot afford to comply with the NPDWR for which a small system variance is sought, including compliance through [SDWA §1415(e)(3) and 40 CFR 142.306(b)(2)]: • Treatment • Alternate source of water supply • Restructuring or consolidation changes • Financial assistance
Ensure Adequate Protection of Human Health	The terms of the small system variance must ensure adequate protection of human health given source water quality, removal efficiencies, and the expected useful life of the SSVT [SDWA §1415(e)(3)(B) and 40 CFR 142.306(b)(5)].

Table continued next page.

Small System Variances (continued)	
Compliance Requirements	
Compliance Date	Systems must comply with the terms of the small system variance within 3 years, unless the state allows up to an additional 2 years to make capital improvements. The state must review each variance at least once every 5 years to determine whether the system remains eligible [SDWA §1415(e)(4)&(5) and 40 CFR 142.307(c)(4)&(d)].
Technology Improvements	Systems must install an SSVT no later than 3 years (with a possible 2-year extension period) after the issuance of the variance and must be financially and technically capable of installing, operating, and maintaining the SSVT [40 CFR 142.306(b)(3)&(4)].
Public Hearing	Before a small system variance may take effect, the state must work with the system to provide public notice to everyone served by the system. Public notice must be issued 15 days before the proposed effective date and 30 days prior to a public meeting [40 CFR 142.308(a)].
Public Notification	Systems must provide public notice within 1 year after the system begins operating under a variance and repeat the notice annually for the duration of the small system variance [40 CFR 141.204(b)(1)].

Exemptions	
Eligibility Requirements	
No Alternative Water Source	The system is unable to comply with the NPDWR due to compelling factors (which may include economic factors) or to implement measures to develop an alternative source of water supply to achieve compliance [SDWA §1416(a)(1) and 40 CFR 142.50(a)(1)].
Does Not Pose an URTH	The state must make a determination that the exemption will not pose an URTH and may require interim compliance measures [SDWA §1416(a)(3) and 40 CFR 142.50(a)(3)].
System Operation	Systems must have begun operation prior to the effective date of the NPDWR; however, this requirement may be waived if the system does not have an alternative source of water supply [SDWA §1416(a)(2) and 40 CFR 142.50(a)(2)].
Management or Restructuring Changes	The system cannot reasonably make management or restructuring changes that would result in compliance or improved quality of the drinking water [SDWA §1416(a)(4) and 40 CFR 142.50(a)(4)].
Unable to Achieve Compliance	No exemption shall be granted unless [SDWA §1416(b)(2)(B) and 40 CFR 142.50(b)(1),(2)&(3)]: • Capital improvements cannot be completed before the NPDWR effective date -or- • A system that needs financial assistance has entered into an agreement to obtain that assistance -or- • The system has entered into an enforceable agreement to become part of a regional public water system; and the system is taking all appropriate steps to meet the standard.

Table continued next page.

Exemptions (continued)	
Compliance Requirements	
Duration	Systems must achieve compliance with the MCL as expeditiously as practicable and in accordance with a compliance schedule determined by the state, but no longer than 3 years from the date of issuance [SDWA §1416(b)(2)(A) and 40 CFR 142.56]. Systems serving <3,301 persons may be eligible for an additional one or more 2-year periods, but the total duration of the exemption extensions may not exceed 6 years [SDWA §1416(b)(2)(C) and 40 CFR 142.56].
Public Hearing	Before an exemption can take effect, the state must provide notice and opportunity for a public hearing on the exemption schedule [SDWA §1416(b)(1)(B) and 40 CFR 142.54(a)].
Public Notification	Systems must provide public notice within 1 year after the system begins operating under an exemption and must repeat the notice annually for the duration of the exemption [40 CFR 141.204(b)(1)].

PUBLIC NOTIFICATION RULE

Highlights

- Revises timing and distribution requirements—notice must be provided within 24 hours (Tier 1, instead of 72 hours), 30 days (Tier 2, instead of 14 days), or 1 year (Tier 3, instead of 90 days), based on the potential severity of the situation.
- Expands list of violations and situations requiring immediate notification and broadens applicability of the public notice to other situations.
- Simplifies mandatory health effects language and adds standard language for monitoring violations and for encouraging notice distribution.
- Consolidates public notification requirements previously found in other parts of drinking water regulations.
- Increases primacy agency flexibility.
- Amends CCR regulations to conform to changes made in public notification regulations.

Title

- Revisions to the Public Notification Regulations for Public Water Systems (40 CFR Part 141, subpart Q), published May 4, 2000 (65 FR 25981).

Purpose

- To notify the public any time a water system violates national primary drinking water regulations or has other situations posing a risk to public health.

Effective Date

- The Public Notification Rule was effective as of June 5, 2000. Public water systems (PWSs) in jurisdictions directly implemented by USEPA must have met these revised requirements by Oct. 31, 2000. PWSs in primacy states must have met these revised requirements by May 6, 2002 or when the state adoopted the revised regulations, whichever was sooner.

Applicability

- All PWSs violating national primary drinking water regulations, operating under a variance or exemption, or having other situations posing a risk to public health.

Timing and Distribution

- Notices must sent within 24 hours, 30 days, or 1 year, depending on the tier to which the violation is assigned (see following discussion). The clock for notification starts when the PWS learns of the violation. Notices must be provided to persons served (not just billing customers).

Table continued next page.

Multilingual Requirements
Where the PWS serves a large proportion of non-English speakers, the PWS must provide information in the appropriate language(s) on the importance of the notice or on how to get assistance or a translated copy.
Tier 1 (immediate notice, within 24 hours)
Notice as soon as practical or within 24 hours via radio, TV, hand delivery, posting, or other method specified by primacy agency, along with other methods if needed to reach persons served. PWSs must also initiate consultation with primacy agency within 24 hours. Primacy agency may establish additional requirements during consultation. • Fecal coliform violations; failure to test for fecal coliform after initial total coliform sample tests positive. • Nitrate, nitrite, or total nitrate and nitrite MCL violation; failure to take confirmation sample. • Chlorine dioxide MRDL violation in distribution system; failure to take samples in distribution system when required. • Exceedance of maximum allowable turbidity level, if elevated to Tier 1 by primary agency. • Special notice for NCWSs with nitrate exceedances between 10 mg/L and 20 mg/L, where system is allowed to exceed 10 mg/L by primacy agency. • Waterborne disease outbreak or other waterborne emergency. • Other violations or situations determined by the primacy agency.
Tier 2 (notice as soon as possible, within 30 days)
Notice as soon as practical or within 30 days. Repeat notice every 3 months until violations are resolved. **CWSs**: Notice via mail or direct delivery. **NCWSs**: Notice via posting, direct delivery, or mail. Primacy agencies may permit alternate methods. All PWSs must use additional delivery methods reasonably calculated to reach other consumers not notified by the first method. • All MCL, MRDL, and treatment technique violations, except where Tier 1 notice is required. • Monitoring violations, if elevated to Tier 2 by primacy agency. • Failure to comply with variance and exemption conditions. **Turbidity consultation:** PWSs that have a treatment technique violation resulting from a single exceedance of the maximum allowable turbidity limit or an MCL violation resulting from an exceedance of the 2-day turbidity limit must consult their primacy agency within 24 hours. Primacy agencies will then determine whether a Tier 1 notice is necessary. If consultation does not occur within 24 hours, violations are automatically elevated to Tier 1.
Tier 3 (annual notice)
Notice within 12 months; repeated annually for unresolved violations. Notices for individual violations can be combined into an annual notice (including the CCR, if public notification requirements can still be met). **CWSs:** Notice via mail or direct delivery. **NCWSs:** Notice via posting, direct delivery, or mail. Primacy agencies may permit alternative methods. All PWSs must use additional delivery methods reasonably calculated to reach other consumers not notified by the first method. • Monitoring or testing procedure violations, unless primacy agency elevates to Tier 2. • Operations under a variance and exemption. • Special public notices (fluoride secondary maximum contaminant level [SMCL] exceedance, availability of unregulated contaminant monitoring results).

Requirements for Ongoing Violations

- All new billing units and customers must be notified of ongoing violations or situations requiring notice.

Relationship to the CCR

Where appropriate, the public notification and CCR requirements are consistent:

- Health effects language for MCL, MRDL, and treatment technique violations are the same.
- Multilingual and certification requirements are similar.
- CCR may be used for Tier 3 notification, provided public notification timing, content, and delivery requirements are met.

Reporting and Record Keeping

- PWSs have 10 days to send a certification of compliance and a copy of the completed notice to the primacy agency.
- PWS and primacy agency must keep notices on file for 3 years.
- Primacy agencies must report public notification violations to USEPA on a quarterly basis.

Primacy Requirements

- Primacy agencies must submit complete and final requests for approval of program revisions in order to maintain primacy for public notification.
- Primacy agencies have up to 2 years to adopt the new regulations.
- Primacy agencies must establish enforceable requirements and procedures if they choose to use any of the flexibilities allowed them in the public notification regulation (e.g., if they allow a PWS to use a different notification method or if they elevate a Tier 2 violation to Tier 1).

Materials Available to Support this Rule

- USEPA/ASDWA *Public Notification Handbook* provides sample notice templates for water systems and other aids for water systems preparing notices.
- *Primacy Guidance for the Public Notification Rule* provides guidance and formats for states preparing primacy program revisions to adopt the public notification rule.

For More Information

- Call the Safe Drinking Water Hotline, at 1-800-426-4791, or access the Office of Ground Water and Drinking Water web site: http://www.epa.gov/safewater/pn.html.

Contents of Notice

Unless otherwise specified in the regulations, each notice must contain:

1. A description of the violation or situation, including contaminant levels, if applicable.
2. When the violation or situation occurred.
3. Any potential adverse health effects (using standard health effects language from Appendix B of the Public Notification Rule or the standard monitoring language, see following).
4. The population at risk.
5. Whether alternative water supplies should be used.
6. What actions consumers should take.
7. What the system is doing to correct the violation or substation.
8. When the water system expects of return to compliance or resolve the situation.
9. The name, business address, and phone number of the water system owner or operator.
10. A statement (see following) encouraging distribution of the notice to others, where applicable.

Standard Language

Standard monitoring language: We are requried to monitor your drinking water for specific contaminants on a regular basis. Results of regular monitoring are an indicator of whether or not our drinking water meets health standards. During [period] we [did not monitor or test/did not complete all monitoring or testing] for [contaminant(s)] and therefore cannot be sure of the quality of the drinking water during that time.

Standard distribution language: Please share this information with all the people who drink this water, especially those who may not have received this notice directly (for example, people in apartments, nursing homes, schools, and businesses). You can do this by posting this notice in a public place or distributing copies by hand or mail.

CONSUMER CONFIDENCE REPORT RULE

Overview	
Title	Consumer Confidence Report (CCR) Rule: 40 CFR, Part 141, Subpart O.
Purpose	Improve public health protection by providing educational material to allow consumers to make educated decisions regarding any potential health risks pertaining to the quality, treatment, and management of their drinking water supply.
General Description	The CCR Rule requires all community water systems to prepare and distribute a brief annual water quality report summarizing information regarding source, any detected contaminants, compliance, and educational information.
Utilities Covered	CWSs, all size categories.

Public Health Related Benefits	
Implementation of the CCR Rule will result in...	• Increased consumer knowledge of drinking water quality, sources, susceptibility, treatment, and drinking water supply management. • Increased awareness of consumers to potential health risks, so they may make informed decisions to reduce those risks, including taking steps toward protecting their water supply. • Increased dialogue with drinking water utilities and increased understanding of consumers to take steps toward active participation in decisions that affect public health.

Annual Requirements	
CWSs with 15 or more connections or serving at least 25 year-round residents must prepare and distribute a CCR to all billing units or service connections.	• *April 1*—Deadline for CWS that sells water to another CWS to deliver the information necessary for the buyer CWS to prepare their CCR (req. outlined in 40 CFR 141.152). • *July 1*—Deadline for annual distribution of CCR to customers and state or local primacy agency for report covering January 1–December 31 of previous calendar year. • *October 1* (or 90 days after distribution of CCR to customers, whichever is first)—Deadline for annual submission of proof of distribution to state or local primacy agency. • A system serving 100,000 or more persons must also post its current year's report on a publicly accessible site on the Internet. Many systems choose to post their reports at the following USEPA web site: http://yosemite.epa.gov/ogwdw/ccr.nsf/america. • All systems must make copies of the report available on request.

Small Water System Flexibility
• With the permission of the governor of a state (or designee) or where the tribe has primacy, in lieu of mailing, systems serving fewer than 10,000 persons may publish their CCR in a local newspaper.* • With the permission of the governor of a state (or designee) or where the tribe has primacy, in lieu of mailing and/or publication, systems serving 500 or fewer persons may provide a notice stating the report is available on request.
*Questions regarding whether the necessary permission has been granted should be addressed to the local state or primacy agency.

Major Provisions to be Included in the CCR
Water System Information
• Name/phone number of contact person. • Information on public participation opportunities (time and place for meetings or hearings). • Information for non-English speaking populations (if applicable).
Source of Water
• Type (e.g., groundwater or surface water), commonly used name, and location of water sources (e.g., Potomac River, Snake River Plain Aquifer, etc.) (Exact locations/coordinates of wells and intakes should not be included for security reasons.) • Availability of source water assessment. • Brief summary on potential sources of contamination (if available).
Definitions
• Maximum contaminant level (MCL) • Maximum contaminant level goal (MCLG) • Treatment technique (TT) (if applicable) • Maximum residual disinfectant level (MRDL) (if applicable) • Maximum residual disinfectant level goal (MRDLG) (if applicable) • Action level (AL) (if applicable) • Variances and exemptions (if applicable)
Detected Contaminants
• Table summarizing data on detected regulated and unregulated contaminants that were detected during the last round of sampling. • Known or likely source of each detected contaminant. • Health effects language for any violations, or exceedances or when arsenic levels are >0.01 mg/L or ≤0.05 mg/L. • Information on *Cryptosporidium*, radon, and other contaminants (if applicable).

Table continued next page.

Major Provisions to be Included in the CCR (continued)
Compliance With Drinking Water Regulations
• Explanation of violations, length of violations, potential health effects, and steps taken to correct the violations. • Explanation of variance/exemption (if applicable).
Required Educational Information
• Explanation of contaminants and their presence in drinking water, including bottled water. • Warning about *Cryptosporidium* for vulnerable or immunocompromised populations. • Informational statements on arsenic, nitrate, lead, and TTHM (if applicable). • USEPA's Safe Drinking Water Hotline number (1-800-426-4791).

STANDARDIZED MONITORING FRAMEWORK FOR CHEMICAL CONTAMINANTS

Overview	
Title	Standardized Monitoring Framework (SMF), promulgated in the Phase II Rule on Jan. 30, 1991 (56 FR 3526).
Purpose	To standardize, simplify, and consolidate monitoring requirements across contaminant groups. The SMF increases public health protection by simplifying monitoring plans and synchronizing monitoring schedules, leading to increased compliance with monitoring requirements.
General Description	The SMF reduces the variability within monitoring requirements for chemical and radiological contaminants across system sizes and types.

Additional Requirements

- The SMF outlined on these pages summarizes existing systems' ongoing federal monitoring requirements only. Primacy agencies have the flexibility to issue waivers, with USEPA approval, which take into account regional and state-specific characteristics and concerns. To determine exact monitoring requirements, the SMF must be used in conjunction with any USEPA-approved waiver and additional requirements as determined by the primacy agency.
- New water systems may have different and additional requirements as determined by the primacy agency.

SMF Benefits

Implementation of the SMF results in…

- Increased public health protection through monitoring consistency.
- A reduction in the complexity of water quality monitoring from a technical and managerial perspective for both primacy agencies and water systems.
- Equalizing of resource expenditures for monitoring and vulnerability assessments.
- Increased water system compliance with monitoring requirements.

Regulated Contaminants	
IOCs	15 (Nitrate, nitrite, total nitrate/nitrite, and asbestos are exceptions to SMF)
SOCs and VOCs	51
Radionuclides	4

Utilities Covered	
All public water systems	Nitrate Nitrite
Community water systems	IOCs SOCs VOCs Radionuclides
Nontransient, noncommunity water systems	IOCs SOCs VOCs

Standardized Monitoring Framework

IOCs, SOCs, VOCs		Second Cycle									Third Cycle								
		1st Period			2nd Period			3rd Period			1st Period			2nd Period			3rd Period		
Inorganic Contaminants (IOCs)[1]	**Groundwater (Below MCL)**	2002	2003	2004	2005	2006	2007	2008	2009	2010	2011	2012	2013	2014	2015	2016	2017	2018	2019
	Waiver [2]					*									*				
	No Waiver		*			*			*			*			*			*	
	Surface Water (Below MCL)																		
	Waiver [2]					*									*				
	No Waiver	*	*	*	*	*	*	*	*	*	*	*	*	*	*	*	*	*	*
	Groundwater and Surface Water (Above MCL)[3]																		
	Reliably and Consistently ≤MCL for Groundwater Systems		*			*			*			*			*			*	
	Reliably and Consistently ≤MCL for Surface Water Systems	*	*	*	*	*	*	*	*	*	*	*	*	*	*	*	*	*	*
	>MCL or Not Reliably and Consistently ≤MCL	****	****	****	****	****	****	****	****	****	****	****	****	****	****	****	****	****	****
Synthetic Organic Contaminants (SOCs)	**Population >3,300 (Below Detection Limit)**	2002	2003	2004	2005	2006	2007	2008	2009	2010	2011	2012	2013	2014	2015	2016	2017	2018	2019
	Waiver		X			X			X			X			X			X	
	<Detect and No Waiver		**			**			**			**			**			**	
	Population ≤3,300 (Below Detection Limit)																		
	Waiver		X			X			X			X			X			X	
	<Detect and No Waiver		*			*			*			*			*			*	
	Above Detection Limit																		
	Reliably and Consistently <MCL[4]	*	*	*	*	*	*	*	*	*	*	*	*	*	*	*	*	*	*
	≥Detect or Not Reliably and Consistently ≤MCL	****	****	****	****	****	****	****	****	****	****	****	****	****	****	****	****	****	****

Table continued next page.

Standardized Monitoring Framework (continued)																			
	Exceptions	Second Cycle									Third Cycle								
		1st Period			2nd Period			3rd Period			1st Period			2nd Period			3rd Period		
Volatile Organic Contaminants (VOCs)	**Groundwater (Below Detection Limit)**	2002	2003	2004	2005	2006	2007	2008	2009	2010	2011	2012	2013	2014	2015	2016	2017	2018	2019
	<Detect, Vulnerability Assessment, and Waiver[5]	*						*						*					
	No Waiver[6]	*	*	*	*	*	*	*	*	*	*	*	*	*	*	*	*	*	*
	Surface Water (Below Detection Limit)																		
	<Detect, Vulnerability Assessment, and Waiver[7]	X			X			X			X			X			X		
	No Waiver[8]	*	*	*	*	*	*	*	*	*	*	*	*	*	*	*	*	*	*
	Above Detection Limit																		
	Reliably and Consistently <MCL[4]	*	*	*	*	*	*	*	*	*	*	*	*	*	*	*	*	*	*
	≥Detect or Not Reliably and Consistently ≤MCL	****	****	****	****	****	****	****	****	****	****	****	****	****	****	****	****	****	****
Nitrate	**CWSs and NTNCWSs**	2002	2003	2004	2005	2006	2007	2008	2009	2010	2011	2012	2013	2014	2015	2016	2017	2018	2019
	Surface Water With 4 Quarters of Results ≤½ MCL[9]	*	*	*	*	*	*	*	*	*	*	*	*	*	*	*	*	*	*
	Groundwater Reliably and Consistently >MCL[9]	*	*	*	*	*	*	*	*	*	*	*	*	*	*	*	*	*	*
	≥½ MCL	****	****	****	****	****	****	****	****	****	****	****	****	****	****	****	****	****	****
	TNCWSs																		
	Standard Monitoring	*	*	*	*	*	*	*	*	*	*	*	*	*	*	*	*	*	*
Nitrite		2002	2003	2004	2005	2006	2007	2008	2009	2010	2011	2012	2013	2014	2015	2016	2017	2018	2019
	<½ MCL					#									#				
	Reliably and Consistently <MCL[9]	*	*	*	*	*	*	*	*	*	*	*	*	*	*	*	*	*	*
	≥½ MCL or Not Reliably and Consistently >MCL	****	****	****	****	****	****	****	****	****	****	****	****	****	****	****	****	****	****

Table continued next page.

Standardized Monitoring Framework (continued)

Exceptions		Second Cycle									Third Cycle								
		1st Period			2nd Period			3rd Period			1st Period			2nd Period			3rd Period		
Radionuclides		2002	2003	2004	2005	2006	2007	2008	2009	2010	2011	2012	2013	2014	2015	2016	2017	2018	2019
	<Detection Limit	!		****				*									*		
	≥Detection Limit but ≤½ MCL			****				*						*					
	>½ MCL but ≤MCL			****				*			*			*			*		
	>MCL			****	****	****	****	****	****	****	****	****	****	****	****	****	****	****	****
Asbestos		2002	2003	2004	2005	2006	2007	2008	2009	2010	2011	2012	2013	2014	2015	2016	2017	2018	2019
	Waiver	X			X			X			X			X			X		
	No Waiver, Reliably and Consistently ≤MCL, or Vulnerable to Asbestos Contamination[10]	*									*								
	<MCL	****	****	****	****	****	****	****	****	****	****	****	****	****	****	****	****	****	****

LEGEND:

- * = 1 sample at each entry point to distribution system (EPTDS).
- ** = 2 quarterly samples at each EPTDS. Samples must be taken during one calendar year during each 3-year compliance period.
- **** = 4 quarterly samples at each EPTDS within time frame designated by the primacy agency.
- X = No sampling required unless required by the primacy agency.
- # = Systems must monitor at a frequency specified by the primacy agency.
- ! = When allowed by the primacy agency, data collected between June 2000 and Dec. 8, 2003, may be grandfathered to satisfy the initial monitoring requirements due in 2004 for gross alpha, radium-226/228, and uranium.

Table continued next page.

Standardized Monitoring Framework (continued)

NOTE:

[1] Until Jan. 22, 2006, the maximum contaminant level (MCL) for arsenic is 50 μg/L; on Jan. 23, 2006, the MCL for arsenic becomes 10 μg/L.

[2] Based on three rounds of monitoring at each EPTDS with all analytical results below the MCL. Waivers are not permitted under the current arsenic requirements; however, systems are eligible for arsenic waivers after Jan. 23, 2006.

[3] A system with a sampling point result above the MCL must collect quarterly samples, at that sampling point, until the system is determined by the primacy agency to be reliably and consistently below the MCL.

[4] Samples must be taken during the quarter that previously resulted in the highest analytical result. Systems can apply for a waiver after three consecutive annual sampling results are below the detection limit.

[5] Groundwater systems must update their vulnerability assessments during the time the waiver is effective. Primacy agencies must reconfirm that the system is nonvulnerable within 3 years of the initial determination or the system must return to annual sampling.

[6] If all monitoring results during initial quarterly monitoring are less than the detection limit, the system can take annual samples. If after a minimum of 3 years of annual sampling with all analytical results less than the detection limit, the primacy agency can allow a system to take one sample during each compliance period. Systems are also eligible for a waiver.

[7] Primacy agencies must determine that a surface water system is nonvulnerable based on a vulnerability assessment during each compliance period or the system must return to annual sampling.

[8] If all monitoring results during initial quarterly monitoring are less than the detection limit, the system can take annual samples. Systems are also eligible for a waiver.

[9] Samples must be taken during the quarter that previously resulted in the highest analytical result.

[10] Systems are required to monitor for asbestos during the first 3-year compliance period of each 9-year compliance cycle. A system vulnerable to asbestos contamination due solely to corrosion of asbestos-cement pipe must take one sample at a tap served by that pipe. A system vulnerable to asbestos contamination at the source must sample at each EPTDS.

Water Quality Basics

Water supplied to customers for human consumption should meet the regulated levels for all contaminants and should be free from aesthetically objectionable characteristics. The following material highlights important water quality characteristics under four general headings: physical, chemical, biological, and radiological. Some treatment methods that a small utility might use to reduce contaminants to acceptable levels are also mentioned.

PHYSICAL PARAMETERS

Physical characteristics of water include turbidity, color, taste and odor, temperature, and foamability. Only turbidity is regulated under the Safe Drinking Water Act (SDWA) and only for surface water sources.

Turbidity

The presence of suspended material in water, such as finely divided organic material, plankton, clay, silt, and other inorganic material, is indicated by the parameter known as turbidity. Turbidity is a measure of the extent to which water scatters light. Turbidity in excess of 5 ntu (nephelometric turbidity units) is easily detected in a glass of water and usually objectionable for aesthetic reasons. More important, high turbidity interferes with disinfection. Turbidity in water from surface water sources, which must always be disinfected, is currently regulated under the SDWA to less than 1 ntu and in some cases even lower.

Clay or other inert suspended particles in water drawn from groundwater sources may not adversely affect health, but water containing these particles may require treatment to make the water aesthetically suitable for its intended use. Following a rainfall, variations in groundwater turbidity may be considered an indication of surface or other introduced pollution.

Turbidity can be removed by filtration. In waters with excessive turbidity, coagulation, flocculation, and sedimentation also may be required. These processes help ensure adequate removal of contaminants and when properly used, reduce loading on filters.

Color

Dissolved organic material from decaying vegetation and certain inorganic matter can cause color in water. Occasionally, excessive algal blooms or growth of aquatic microorganisms may also cause color. Although color alone is not usually objectionable from a health standpoint, its presence is often aesthetically objectionable. In addition, when color is produced from organic molecules, a reaction with chlorine may cause the formation of trihalomethanes (THMs). These and other chlorinated organics have been identified as potential carcinogens and are regulated by the SDWA. See the section on organic chemicals later in this chapter for more information.

Color molecules are small, and a combination of treatment techniques, including coagulation, flocculation, sedimentation, and filtration, are required to reduce color to an acceptable level. Other methods for reducing levels of color include chemical oxidation using potassium permanganate, chlorine dioxide, or ozone; membrane filtration; and application of granular activated carbon (GAC).

Taste and Odor

Taste and odor in water can be caused by organic compounds, inorganic salts, or dissolved gases and may come from domestic, agricultural, synthetic, or natural sources. Water should be free from any objectionable tastes or odors before it flows into the distribution system. Aeration, filtration, chemical oxidation, and adsorption with activated carbon can remove tastes and odors in treatment systems. In some cases, pretreatment at the water source may eliminate the cause of tastes and odors.

Temperature

The most desirable drinking water is consistently cool with temperature fluctuations of less than a few degrees. Groundwater and surface water from mountainous areas generally meet these criteria. Most individuals find that water between 50°F and 60°F (10°C and 15°C) is most palatable.

Foamability

Foaming in water is usually caused by detergent concentrations greater than 1 mg/L. Foam alone may not be hazardous, but it is objectionable and should not be tolerated. Water with high foamability should be analyzed to determine potential sources of the problem as well as what treatment may be required. Results of sanitary surveys can be used to determine the origin of contamination and the opportunities for better source protection. Adsorption with activated carbon generally removes foaming agents.

CHEMICAL PARAMETERS

The soil overburden—rocks and bedrock—that forms the earth's crust has a direct effect on the quality and quantity of groundwater that may be withdrawn. As surface water seeps into the water table, it often dissolves minerals contained in the soils and rocks. Groundwater, therefore, often contains more dissolved minerals than does surface water. Objectionable synthetic chemicals may be found in both surface supplies and groundwater.

The chemical characteristics of a water source should be analyzed with special attention to the:

- possible presence of any harmful or disagreeable substances,
- potential for water to corrode parts of the water system, and
- potential for water to stain plumbing fixtures and clothing.

Chemical characteristics that may be identified during a water analysis are discussed in the following paragraphs. The sample size and method of collection should be in accordance with recommendations of the analytical laboratory.

Alkalinity

Alkalinity is imparted to water by bicarbonate, carbonate, and hydroxide components. The presence of these components is determined using standard titration methods with various indicator solutions. Knowledge of alkalinity components is useful in treating water supplies as a measurement of the water's ability to neutralize acidity.

Chloride: secondary MCL = 250 mg/L

Most waters contain some chloride caused naturally through leaching of marine sedimentary deposits within the bedrock and by pollution from seawater, brine, road salting, or industrial and domestic wastes. Chloride concentrations in excess of 250 mg/L (the MCL) usually produce a noticeable salty taste in drinking water. Where chloride content is higher than 250 mg/L but the water meets all other criteria, it may be necessary to use a water source that exceeds this limit for chloride. High chloride concentrations contribute to the corrosiveness of water when in contact with metallic pipes and heating equipment. In some cases, high chloride is accompanied by high sodium, which can be a health concern to some individuals. Chloride ions are very small. If the only source of water available has high chloride levels, it may be necessary to use reverse osmosis (RO), another membrane technology, or distillation units to produce potable water.

Copper: action level for corrosion control = 1.3 mg/L

The action level is the concentration of copper in water that determines the corrosion control treatment requirements for a water system. Copper is found in some natural waters, particularly where copper has been mined. Although everyone needs copper in small amounts, excessive amounts of copper can cause acute toxic effects. Copper leaching occurs when corrosive water passes through unprotected copper pipes. This can cause taste problems, stain fixtures, and cause chemical interactions for people with tinted hair. Corrosion from copper can be eliminated by adjusting the water's pH to decrease its corrosivity. Copper is rarely found in naturally occurring waters at levels that require treatment, but it can be removed by conventional coagulation, sedimentation, and filtration; softening; or RO.

Corrosivity: secondardy MCL = noncorrosive

The tendency of water to corrode pipes and fittings is a measure of the water's corrosivity. This can be a health-related issue, as well as economically important. Corrosive water may leach into solution materials, including lead, cadmium, copper, chromium, and other toxic metals. Water's corrosivity cannot be measured simply. It may be necessary to review a number of factors to predict the treatment best suited to lowering corrosivity. These factors include temperature, total dissolved solids (TDS), calcium content, pH, and alkalinity. Equations have been developed to predict the calcium carbonate stability of water with reasonable levels of alkalinity and hardness. These equations indicate a water's tendency to either deposit or dissolve calcium carbonate ($CaCO_3$). In most cases, a water that is neutral or slightly scale-forming is preferred. It may be best to treat waters with low alkalinity, low pH, and low dissolved oxygen (DO) using aeration, pH adjustment, polyphosphates, and silicates or some combination of these processes. Local conditions vary, and state or local health departments or nearby utilities using water of similar quality should be consulted for their experience in dealing with corrosion control.

Fluoride: MCL = 4.0 mg/L, secondary MCL = 2.0 mg/L, temperature variable

Where fluoride occurs naturally within a range of <1 mg/L, the incidence of dental caries is below that for areas without fluoride. The improved dental effect is the same whether fluoride occurs naturally or is added during treatment. The optimal fluoride level for a given area depends on air temperature because temperature is the primary influence on the amount of water people drink (see Table 3-1). Concentrations from 0.7 to 1.2 mg/L are recommended.

The fluoride regulation sets both the maximum contaminant level goal (MCLG) and MCL at 4.0 mg/L to protect against crippling sketetal fluorosis. A secondary MCL is set at 2.0 mg/L to protect against objectionable dental fluorosis, not considered by USEPA to be an adverse health effect. State and local health departments should be consulted for recommendations. Fluoride removal is difficult because of the small size of the molecule. RO, alone or in combination with other membrane processes, may be used, along with some specific ion exchange resin applications for smaller treatment systems. In larger volumes, treatment using activated alumina has been successful. This is a relatively complex process if the activated alumina is to be backwashed and reused rather than discarded after one pass.

Table 3-1 Optimal fluoride concentrations

Annual Average of Maximum Daily Air Temperature*		Recommended Control Limits of Fluoride Concentration, mg/L		
°F	°C	Lower	Optimal	Upper
53.7 and below	12.0 and below	0.9	1.2	1.7
53.8–58.3	12.1–14.6	0.8	1.1	1.5
58.4–63.8	14.7–17.6	0.8	1.0	1.3
63.9–70.6	17.7–21.4	0.7	0.9	1.2
70.7–79.2	21.5–26.2	0.7	0.8	1.0
79.3–90.5	26.3–32.5	0.6	0.7	0.8

*Based on temperature data for a minimum of 5 years.

Hardness: low, 0–75 mg/L; moderate, 75–150 mg/L; hard, 150–250 mg/L; very hard, 250 mg/L

Hardness is generally caused by the presence of calcium and magnesium ions in water. Hard water retards the cleaning action of soaps, raising costs for labor and cleaning agents. Synthetic detergents are less affected by hardness. When hard water is heated, it deposits a scale on heating coils, cooking utensils, and other equipment, because minerals that cause hardness are *less* soluble in hot water. Over time, scale formed by hard water coats the inside of distribution system piping, contributing to tuberculation and eventually significantly reducing water mains' carrying capacity. Water with little or no hardness—soft water—can also be corrosive, as discussed previously.

Hardness of 75 to 150 mg/L as $CaCO_3$ is usually considered optimal for domestic water. Water harder than 250 mg/L as $CaCO_3$ should be considered for treatment. Water softer than 30 mg/L as $CaCO_3$ often causes corrosion problems.

Calcium and magnesium salts, the most common cause of hardness in water supplies, are divided into two general classifications: carbonate, or temporary, hardness and noncarbonate, or permanent, hardness. Carbonate hardness is also called temporary hardness because heating the water usually removes the hardness. When water is heated, bicarbonates break down into insoluble carbonates that precipitate as solid particles and adhere to heated surfaces and inside pipes. Noncarbonate hardness is called permanent hardness because it is not removed when water is heated. Noncarbonate hardness is caused largely by the presence of sulfates and chlorides of calcium and magnesium in the water.

Chemical (lime-soda ash) or ion exchange softening processes can produce acceptably soft water where only excessively hard water is available. Membrane filtration may reduce hardness to acceptable levels in small-scale applications. If hardness is lowered too much during any softening process, corrosion problems can occur unless further treatment stabilizes the water.

Iron: secondardy MCL = 0.3 mg/L

Iron is classified as a secondary contaminant, meaning it may adversely affect the odor, taste, or appearance of drinking water. Small amounts of iron are frequently present in water because iron is a natural part of rocks and soils. Iron is generally dissolved and colorless in soil; when it comes in contact with oxygen, it forms solid reddish-brown particles. Iron is also a principal component of piping material and subject to corrosion. Various methods are available for removing iron. In large-scale applications, oxidation, conventional coagulation, sedimentation, and filtration are generally effective. Chemical oxidation, or aeration, followed by filtration is effective, as is cationic exchange resin softening.

Lead: action level for corrosion control = 0.015 mg/L

High exposure to lead has long been recognized as a cause of acute adverse health effects. High lead levels in the blood can cause anemia, kidney damage, and physical and mental retardation. Lead occurs in drinking water primarily as a result of corrosion of pipes, solder, and faucets. The goal of the SDWA's Lead and Copper Rule is to minimize exposure in drinking water; the MCLG is set at zero. The action level is the concentration of lead in water that determines the corrosion control treatment requirements for a system. Treatment requirements consist of optimal corrosion control, source water treatment, public education, and lead service line replacement.

Manganese: secondary MCL = 0.05 mg/L

The chemistry of manganese is similar to that of iron. When oxidized, manganese causes a brownish to black discoloration; at elevated levels, it may also produce taste and odor. There is some evidence of possible physiological effects, primarily heart problems, from excessive manganese. Essentially the same treatment processes used to remove iron may be used to reduce manganese levels. However, manganese is harder to remove than iron because its precipitation is more pH dependent and contact time is longer before oxidation occurs.

Nitrate: MCL = 10.0 mg/L; Nitrite: MCL = 1.0 mg/L (both measured as nitrogen)

Elevated levels of nitrate (NO_3) can cause methemoglobinemia (infant cyanosis, or "blue baby" disease). This condition also affects older people. When ingested, nitrate is converted (reduced) to nitrite, which is a health hazard. Although nitrates can occur naturally, excess concentrations may indicate pollution from fertilizer, sewage, septic system leaching, or manure deposits. In some polluted waters, nitrite is present in concentrations greater than 1 mg/L. Nitrite is more hazardous to infants than nitrate. Nitrates can be removed using an anionic ion exchange process. For small-scale applications, RO units and other membrane technologies can also be used.

Organic Chemicals

Organic chemicals constitute a wide variety of potential contaminants, including pesticides, herbicides, insecticides, THMs (chlorinated organics), volatile organic compounds, and synthetic organic compounds. Careless use or disposal of many synthetic chemicals may cause long-term pollution of water resources. Although many of these compounds are volatile and evaporate into the open air, they remain unaffected and unchanged for long periods of time if they are introduced into groundwater. These chemicals should not be used or disposed of near water supplies. The SDWA sets MCLs for these chemicals.

THMs are a group of organic compounds that form when chlorine reacts with humic and fulvic acids (natural organic compounds that occur in decaying vegetation). Total trihalomethanes (TTHMs), a combination of all chlorinated organic molecules, are considered potential carcinogens (cancer-causing agents) and should not exceed 0.1 mg/L in drinking water. Haloacetic acids (HAAs) are a family of disinfection by-products (DBPs) to be regulated by the US Environmental Protection Agency (USEPA) under the SDWA.

Volatile organic compounds and synthetic organic compounds occur as waste products of various industrial and commercial processes and have been found in groundwater near industrial and commercial areas. At elevated levels these chemicals have toxic effects; at trace levels they are suspected of being carcinogenic. Organic chemicals can generally be removed by adsorption with activated carbon. THMs can often be prevented by removing organics before chlorination, altering the chlorination application point, or using an alternate disinfectant such as chloramines, chlorine dioxide, or potassium permanganate. Most volatile organic compounds can be eliminated by aeration. Protecting the water source from this type of pollution is a primary obligation for the drinking water community.

pH: secondary MCL = 6.5–8.5

pH is defined as the negative log of the hydrogen ion concentration in water. It is a measure of the water's acidic or alkaline content. pH values range from 0 to 14, with 7 indicating neutral water. A change of one unit of pH actually reflects a tenfold difference in acid content, because pH units are based on a logarithmic scale. Values less than 7 indicate the water is more acidic; values greater than 7 indicate the water is more alkaline. Normally, the pH of water in its natural state varies from ±5.5 to ±9.0. Determining the pH value assists in corrosion control, especially for copper corrosion. It is also a factor in calculating proper chemical dosages and adequate disinfection.

Sodium: guidance level = 20 mg/L

The guidance level is the level that is recommended for persons on low-sodium diets. A water's sodium content is generally not important to the general population, because people ingest less than 10 percent of their daily intake of sodium via water. Most sodium comes from foods and salt. Certain at-risk groups are on low-sodium diets because of heart, kidney, or circulatory ailments or because of complications of pregnancy. These individuals must be aware of the sodium content of the water they drink. When it is necessary to know the precise amount of sodium present in a water supply, a laboratory analysis should be performed. Most low-sodium diets allow for up to 20 mg/L sodium in drinking water. When this limit is exceeded, those on low-sodium diets may have to drink water from a source other than the public supply. Some water treatment processes, such as ion exchange softening, increase the amount of sodium in drinking water. For this reason, softened water should be analyzed for sodium if a precise record on an individual's sodium intake is recommended. Because sodium is a small ion, reduction is best achieved on a limited basis with RO or distillation units.

Sulfate: secondary MCL = 250 mg/L

Sulfate is found in most natural waters. High concentrations of sulfate in source waters may be caused by natural deposits of magnesium sulfate (Epsom salts) or sodium sulfate (Glauber's salt). High sulfate levels may be undesirable because of its laxative effects. Sulfate may also enter water as a pollutant from septic system leaching. When sulfate and sulfate-reducing bacteria are present together, they produce a by-product—hydrogen sulfide gas (H_2S)—that causes an unpleasant sulfur or "rotten egg" odor. RO, specialized ion exchange, or distillation can be used to reduce sulfate concentrations.

Total Dissolved Solids: secondary MCL = 500 mg/L

Total dissolved solids (TDS) is a measure of various dissolved minerals in water. Water with no TDS, such as distilled water or water treated by RO, usually has a flat taste. Water with more than 500 mg/L TDS usually has a disagreeably strong taste, depending on the solids dissolved in the water. RO, other membrane filtration, softening, or ion exchange may be used to successfully reduce TDS content, depending on the chemical nature of the dissolved solids.

Toxic Metals

Arsenic, barium, cadmium, chromium, mercury, selenium, and silver can all cause serious health problems, even fatalities, if they are present in drinking water in more than trace amounts. The principal problem with heavy metals is that they accumulate in the human body. Once ingested, metals are retained in body tissue and, over time, the total retained amount can become toxic. Softening processes, ion exchange, precipitation with alum, activated alumina filtration, and RO can reduce the concentrations of toxic metals.

Zinc: secondary MCL = 5 mg/L

Zinc is found in some natural waters, particularly where zinc has been mined. Although it is not considered detrimental to health, zinc does impart an undesirable taste. Corrosive waters can also dissolve zinc from galvanized plumbing fixtures into the drinking water. Softening, RO, and ion exchange reduces zinc concentrations. Noncorrosive water in the distribution system limits problems from galvanized plumbing.

BIOLOGICAL PARAMETERS

Water for drinking and cooking purposes must be free from organisms that might cause illness. Organisms commonly found in source waters (primarily surface waters) include bacteria, protozoa, viruses, and worms.

Bacteriological Quality

Specific disease-causing organisms present in water are difficult to identify. Although there are techniques for performing a comprehensive bacteriological examination of water, these techniques are generally organism specific, complex, and time-consuming. The relative degree of bacteriological contamination can be found using a simple, accurate, easily performed test that can provide results within one day's time.

Many microorganisms that cause disease in humans are transmitted through fecal wastes of infected individuals, including *Giardia* cysts and *Cryptosporidium* oocysts. Consequently, the most widely used method of testing water's bacteriological quality involves testing for a single group of bacteria—the coliform group—that is always present when fecal contamination exists. Because coliform bacteria normally inhabit the intestinal tract of humans and other warm-blooded mammals, they are referred to as indicator organisms. Coliform bacteria are not harmful to humans, but their presence indicates that other harmful disease-causing microorganisms might also be present—a problem that must be immediately remedied.

The two methods most frequently used to test specifically for coliform are the presence–absence test and the membrane filter test. The presence–absence test is quick, reliable, and easy. Testing results should show an absence of *Escherichia coli,* fecal coliform, and total coliform from treated water before it flows to the distribution system. The membrane filter test discriminates between coliform and noncoliform bacteria. High numbers of noncoliform bacteria, although probably not harmful, indicate a source of pollution in the sample. A third, older, standard test is the multiple-tube fermentation test. Also, the heterotrophic (standard) plate count can provide a general indication of biological organisms by identifying any number of species growing under certain conditions in a water sample.

Protozoa and viruses form two of the three categories for pathogenic organisms that may be waterborne. These organisms are more resistant to chlorine than bacteria are, and their removal relies more on optimizing treatment processes.

Some groundwater sources, if properly protected and developed, can meet these standards without treatment. However, groundwater disinfection is a recommended safeguard and may be required by state or local health agencies, or both. Chlorination of groundwater also introduces a disinfectant residual that helps maintain bacteriological quality of the water in the distribution system.

Under the Surface Water Treatment Rule (SWTR), water from surface sources must be filtered and disinfected. A waiver process is available for surface water filtration, but disinfection is mandatory. Although chlorination is the most common form of disinfection, additional methods for use in small systems include application of other forms of chlorine, such as chlorine dioxide and chloramine; bromine; iodine; ozone; and ultraviolet light (UV). Each method has its advantages and disadvantages. It is essential to disinfect water before it enters the distribution system. The best insurance against bacterial problems other than disinfection is to use source water protection as a barrier to potential contamination. This is a key, ongoing priority for the drinking water community.

Other Biological Factors

Certain forms of aquatic vegetation and microscopic organisms in surface water, although not harmful to humans, may produce by-products that cause unpleasant tastes and odors. Algal blooms in surface waters are a specific example of this type of pollution. Certain nonpathogenic bacteria or microscopic crustacea that inhabit natural surface waters—again, not harmful to humans—may cause similar problems. Copper sulfate, applied in accordance with the recommendations of the state or local health department, is commonly used to eliminate temporarily high algae concentrations in source water reservoirs and lakes.

In groundwater sources, iron bacteria and sulfate-reducing bacteria can cause color, staining, tastes, and odors. Sanitary well-drilling procedures help prevent iron bacteria from entering a new well. Iron bacteria and sulfate-reducing bacteria can usually be eliminated from an existing well by temporarily introducing a high chlorine concentration. If bacteria persist, other methods—such as adding polyphosphates with a high chlorine dosage, recirculating for a period of time, and flushing to open discharge—may be successful.

RADIOLOGICAL PARAMETERS

Humans are continually exposed to natural radiation from sun, water, food, and air. The amount of radiation exposure varies with the amount of normal background radioactivity in the surrounding area.

Radiation from water is caused by naturally occurring radioactive minerals that may be found in water.

Radioactivity may be present from minerals in the environment or, on occasion, from manufactured pollution sources. Human exposure to radiation or radioactive materials increases cancer risks; any unnecessary exposure should be avoided. Concentrations of radioactive materials specified in the current SDWA are intended to limit human intake of radioactive substances. These limits ensure that total radiation exposure of any individual will not exceed the radiation protection guides recommended by the Federal Radiation Council.

Radiological data indicating both background and other forms of radioactivity in an area are available from the USEPA, US Public Health Service, US Geological Survey, and other federal, state, and local agencies. Water from a source under investigation must be tested for radioactivity levels before it is supplied for public consumption.

Softening techniques such as lime softening, aeration, use of GAC or ionic resins, and membrane filtration may remove radioactive minerals. If radioactive minerals are concentrated in media such as granular or powdered activated carbon, the resultant product may be considered a low-level radioactive waste under some conditions, in which case, disposal options may be difficult and expensive.

SANITARY SURVEY

The importance of a comprehensive sanitary survey of potential water sources, both surface water and groundwater, cannot be overemphasized. Even existing sources should be surveyed routinely through a USEPA program; these programs are usually coordinated through the state water supply regulatory agency.

The sanitary survey of a potential supply should be made in conjunction with the collection of initial engineering data regarding the development of a given source and its capacity to meet existing and future needs. Most states require guidance documents for sanitary surveys as part of the approval process for a new water resource. Surveys of the wellhead protection area around groundwater sources and of watersheds around surface supplies are an essential part of the approval process. These surveys cover large geographic areas beyond the immediate area surrounding the actual water supply source. Regulations are more comprehensive and complex for water supply quantities in excess of 70 gpm (4.4 L/s).

The sanitary survey must detect all potential health hazards and assess their present and future importance. This assessment includes known sources of contamination, such as landfills and hazardous waste sites, as well as potential contamination sources, such as fuel storage areas, industrial plants, and sewage treatment facilities. People who are trained and competent in water supply engineering should conduct the sanitary survey. State or health department officials can provide assistance with the survey.

In the case of an existing supply, the survey is generally scheduled every 3 to 5 years by the state; its frequency should be compatible with controlling health hazards and maintaining good sanitary quality. Information furnished by the sanitary survey is essential for complete interpretation of bacteriological and often chemical data. This information should always accompany laboratory findings.

Groundwater Supplies

The following factors relating to groundwater supply should be investigated during a sanitary survey. Not all items are pertinent to every supply, and not all items that would be important additions to a survey are on the list.

- Character of local geology, general slope of ground surface near the source, nature of soil and underlying porous strata—clay, sand, gravel, bedrock (especially porous limestone); coarseness of sand or gravel; thickness of water-bearing stratum; depth to water table; depth to bedrock; location, log, and construction details of local wells in use or abandoned in the local area

- Slope of water table, preferably as determined from observational wells or as indicated presumptively, not by slope of ground surface (best determined during pumping tests or from a review of pumping test records on other nearby wells)

- Extent of drainage area likely to contribute water to the supply, somewhat different from the theoretical wellhead protection area (WHPA) that may be a 1-mile (1.61-km) radius around the source; drainage area varies based on projected well capacity

- Nature, distance, and direction of known and potential contamination sources (PCSs) within the WHPA

- Possibility of surface drainage water entering the supply, potential for wells becoming flooded; likelihood of implementing preventive procedures
- Protecting the supply against potential pollution by zoning and working with state and local regulators for source protection and other programs
- Well construction using a schematic diagram as a well profile identifying:
  - total depth of well
  - casing information, including diameter, wall thickness, material, and total length of casing installed; drive shoe, grouting, or other methods used to ensure a sanitary seal
  - screen or perforations, including diameter, material, installation details, locations, and lengths, or notations from the drilling log on fracture location and approximate yield
  - presence of casing and/or borehole annulus seals, including type of material (cement, sand, bentonite, cuttings), seal interval, method of placement, or other seals associated with telescopic casings and screens with depth of setting and details
- Protection of well at the surface, including presence of sanitary well seal; casing height above ground (at minimum, top of casing should be at a higher elevation than the 100-year floodplain); slope of land immediately around wellhead; condition of screened well vent; protection of well from erosion, animals, and vandals
- Well pump house construction; well pump information including condition of floors, walls, roof, insulation, doors, stairs, drains, piping, valves, master water meter, chemical feed equipment and day tank storage for chemicals, electricity, ventilation, heat, pump capacity, static level, well drawdown or pumping level when pump is in operation, amperage draw and rate of flow during pumpage compared to pump curve data, noting any unusual conditions (such as water hammer) during pump startup and shutdown
- Availability of a backup source of water in the event the normal water supply well is lost, including other wells, emergency connections with adjacent utilities, names and telephone numbers of water-hauling companies
- Emergency disinfection equipment, including method of application or operation, supervision, test kits, and other types of laboratory control, such as bacteria test bottles

Surface Water Supplies

Lakes, rivers, and other surface sources present special challenges when maintaining water quantity, largely due to the variabiity of water quality. The following factors relating to surface supplies should be investigated during a sanitary survey. Not all items are pertinent to every supply, and not all items that would be important additions to a survey are on the list.

- Nature of surface geology, including the character of soils and rocks around the supply source
- Character of vegetation in forested, cultivated, and irrigated land within the watershed; identification of the watershed, including the extent of drainage area likely to contribute water to the supply; nature, distance, and direction of known and potential contamination sources within the general watershed

- Population data, including density, number, and distribution; methods of wastewater disposal for the population within the watershed, noting the character and efficiency of sewage treatment works within the watershed
- Ability of the utility to control the entire watershed through ownership, zoning, easements, and other methods
- Physical and chemical parameters associated with the source noting annual variations in flow, water temperature, and water quality (bacteriological and aquatic, physical, and inorganic parameters); historical records of algae blooms; variations in general water chemistry over the year based on seasonal changes
- Depth of source relative to intake design and location within the water supply, as well as characteristics of the vegetation and bottom characteristics of the source
- Adequacy of supply for long-term sustainable quantity and quality
- Projected nominal detention time and variations of same in the source
  - — probable minimum time required for water to flow from potential sources of pollution, including accidental releases, e.g., a traffic accident on a road close to the source intake
  - — effect of reservoir configuration on possible currents of water induced by wind or reservoir discharge from inlet to water supply intake
  - — watershed protection, including control of fishing, boating, landing of airplanes, swimming, wading, ice cutting, and permitting animals on marginal shore areas and in or on the water
- Efficiency and constancy of policing reservoir and surrounding area
- Type and adequacy of water treatment, adequacy of supervision and testing, contact period after disinfection, and free chlorine residuals
- Pumping facilities, including the pump house, pump capacity and standby units, and storage facilities

Appendix A
Abbreviations and Acronyms

Ag	silver
AIDS	acquired immune deficiency syndrome
AL	action level
As	arsenic
ASDWA	Association of State Drinking Water Administrators
AWWA	American Water Works Association
Ba	barium
BAT	best available technology
CCR	Consumer Confidence Report
CCT	corrosion control treatment
Cd	cadmium
CDC	Centers for Disease Control and Prevention
CFE	combined filter effluent
CFR	Code of Federal Regulations
CPE	comprehensive performance evaluation
Cr^{+6}	chromium
CT	contact time
Cu	copper

CWA	Clean Water Act
CWS	community water system
DBCP	1,2-dibromo-3-chloropropane
DBP	disinfection by-products
DBPP	disinfection by-product precursor
D/DBPR	Disinfectants and Disinfection By-products Rule
DE	diatomaceous earth
DO	dissolved oxygen
DOC	dissolved organic carbon
DWSRF	Drinking Water State Revolving Fund
EPTC	ethyl dipropylthiocarbamate
EPTDS	entry point to distribution system
ESWTR	Enhanced Surface Water Treatment Rule
FACA	Federal Advisory Committee Act
FBRR	Filter Backwash Recycling Rule
FR	*Federal Register*
GAC	granular activated carbon
GAC 10	granular activated carbon with 10-minute empty bed contact time and 180-day reactivation frequency
gpm	gallons per minute
GPO	Government Printing Office
GWR	Ground Water Rule
GWUDI	groundwater under the direct influence of surface water
HAA	haloacetic acid
HAA5	haloacetic acids (five)
Hg	mercury
HPC	heterotrophic plate count
ICR	Information Collection Rule
IDSE	initial distribution system evaluation
IESWTR	Interim Enhanced Surface Water Treatment Rule
IFE	individual filter effluent
IOC	inorganic chemical
LCR	Lead and Copper Rule
LRAA	locational running annual average
L/s	liters per second
LSL	lead service line
LSLR	lead service line replacement
LT1ESWTR	Long-Term 1 Enhanced Surface Water Treatment Rule
LT2ESWTR	Long-Term 2 Enhanced Surface Water Treatment Rule
MCL	maximum contaminant level
MCLG	maximum contaminant level goal
M-DBP	microbial and disinfectants/disinfection by-products
MFL	million fibers per liter

mg/L	milligrams per liter
MRDL	maximum residual disinfectant level
MRDLG	maximum residual disinfectant level goal
mrem	millirem
MTBE	methyl tertiary butyl ether
NCOD	National Contaminant Occurrence Database
NCWS	noncommunity water system
NPDWR	National Primary Drinking Water Regulations
NSDWR	National Secondary Drinking Water Regulations
NTNCWS	nontransient, noncommunity water system
ntu	nephelometric turbidity unit
OGWDW	Office of Ground Water and Drinking Water
OWQP	optimal water quality parameters
PAH	polyaromatic hydrocarbon
Pb	lead
PCB	polychlorinated biphenyl
pCi/L	picocuries per liter
PCS	potential contamination source
PE	public education
pg	picogram
PN	public notification
POE	point of entry
POU	point of use
ppb	parts per billion
ppm	parts per million
PSA	public service announcement
PVC	polyvinyl chloride
PWS	public water system
PWSS	public water supply supervision
reg neg	regulatory negotiation
RO	reverse osmosis
RUS	Rural Utilities Service
SBREFA	Small Business Regulatory Enforcement Fairness Act
SDWA	Safe Drinking Water Act, or the "Act," as amended in 1986 and 1996
Se	selenium
SMCL	secondary maximum contaminant level
SMF	Standardized Monitoring Framework
SOC	synthetic organic chemical
SOWT	source water treatment
SSCT	small system compliance technology
SSVT	small system variance technology
SUVA	specific ultraviolet absorbance
SWTR	Surface Water Treatment Rule

TCR	Total Coliform Rule
TDS	total dissolved solids
THM	trihalomethane
TOC	total organic carbon
TT	treatment technique
TTHM	total trihalomethanes
TNCWS	transient, noncommunity water systems
UCMR	Unregulated Contaminant Monitoring Rule
URTH	unreasonable risk to health
USEPA	United States Environmental Protection Agency
UV	ultraviolet light
VOC	volatile organic chemical
WHPA	wellhead protection area
WQP	water quality parameter

Appendix B
Glossary

These definitions are not intended to be complete or to have legal force but rather to help consumers quickly understand drinking water–related terms in the context of their daily lives.

acidic The condition of water or soil such that it contains a sufficient amount of acidic substances to lower the pH below 7.0.

accuracy The closeness with which an instrument measures the true or actual value of the process variable being measured or sensed. See also *precision*.

action level The level of lead or copper that, if exceeded in more than 10% of homes tested, triggers treatment or other requirements that a water system must follow.

activated carbon Adsorptive particles or granules of carbon usually obtained by heating carbon (such as wood). These particles or granules have a high capacity to selectively remove certain trace and soluble organic materials from water.

acute health effect An immediate (i.e., within hours or days) adverse effect on a person's health that may result from exposure to certain drinking water contaminants (e.g., pathogens).

air binding A situation in which air collects within filter media.

algae Microscopic plants that contain chlorophyll and live floating or suspended in water. They also may be attached to structures, rocks, or other submerged surfaces. They are food for fish and small aquatic animals.

alkaline The condition of water or soil such that it contains a sufficient amount of alkaline substances to raise the pH above 7.0.

alkalinity The capacity of water to neutralize strong acids. This capacity is caused by the water's content of carbonate, bicarbonate, hydroxide, and occasionally, borate, silicate, and phosphate. Alkalinity is expressed in milligrams per liter of equivalent calcium carbonate. Alkalinity is not the same as pH because water does not have to be strongly basic (high pH) to have a high alkalinity. Alkalinity is a measure of how much acid can be added to a liquid without causing a great change in pH.

alternative compliance criteria The eight criteria in the Stage 1 Disinfectants/Disinfection By-products Rule that systems may use to demonstrate compliance with the disinfection by-product precursor (total organic carbon) removal requirements in lieu of the requirement to remove specified levels of disinfection by-product precursors.

aquifer A natural underground layer, often of sand or gravel, that contains water.

available expansion The vertical distance from the filter surface to the overflow level of a trough in a filter. This distance is also called *freeboard.*

backwash The process of reversing the flow of water back through filter media to remove entrapped solids.

bacteria (singular: bacterium) Microscopic living organisms usually consisting of a single cell. Some bacteria in soil, water, and air may also cause human, animal, and plant health problems.

baffle A flat board or plate, deflector, guide, or similar device constructed or placed in flowing water or slurry systems to cause more uniform flow velocities; absorb energy; and divert, guide, or agitate liquids (e.g., water, chemical solutions, slurry).

best available technology (BAT) The water treatment(s) that the US Environmental Protection Agency certifies to be the most effective for removing a contaminant.

breakthrough A condition whereby filter effluent water quality deteriorates (as measured by an increase in turbidity, particle count, or other contaminant). This may occur due to excessive filter run time or hydraulic surge.

calcium carbonate ($CaCO_3$) equivalent An expression of the concentration of specified constituents in water in terms of their equivalent value to calcium carbonate. For example, the hardness in water that is caused by calcium, magnesium, and other ions is usually described as calcium carbonate equivalent.

calibration A procedure used to check or adjust an instrument's accuracy by comparing the instrument's reading with a standard or reference sample that has a known value.

capital costs Costs of construction and equipment. Capital costs are usually fixed, one-time expenses, although they may be paid off over longer periods of time.

carcinogen Any substance that tends to cause cancer in an organism.

chronic health effect The possible result of exposure over many years to a drinking water contaminant at levels above its maximum contaminant level (MCL).

clarifier A large circular or rectangular tank or basin in which water is held for a period of time, during which the heavier suspended solids settle to the bottom by gravity. Clarifiers are also called settling basins or sedimentation basins.

clearwell A reservoir for the storage of filtered water with sufficient capacity to prevent the need to vary the filtration rate in response to short-term changes in customer demand. Also used to provide chlorine contact time for disinfection.

coagulant A chemical added to water that has suspended and colloidal solids to destabilize particles, allowing subsequent floc formation and removal by sedimentation, filtration, or both.

coagulant aid A chemical added during coagulation to improve the process by stimulating floc formation or by strengthening the floc so it holds together better.

coagulation As defined in 40 CFR 141.2, a process using coagulant chemicals and mixing by which colloidal and suspended materials are destabilized and agglomerated into flocs.

cohesion Molecular attraction that holds two particles together.

coliform A group of related bacteria whose presence in drinking water may indicate contamination by disease-causing microorganisms.

colloid A small, discrete solid particle in water that is suspended (not dissolved) and will not settle by gravity because of molecular bombardment.

combined filter effluent Effluent from individual filters that is combined into one stream.

combined sewer A sewer that transports surface runoff and human domestic wastes (sewage) and sometimes industrial wastes.

community water system (CWS) As defined in 40 CFR 141.2, a public water system that serves at least 15 service connections used by year-round residents or regularly serves at least 25 year-round residents.

compliance The act of meeting all state and federal drinking water regulations.

contaminant Anything found in water (including microorganisms, minerals, chemicals, radionuclides, etc.) that may be harmful to human health.

continuous sample A constant flow of water from a particular place in a plant to the location where samples are collected for testing.

conventional filtration treatment As defined in 40 CFR 141.2, a series of processes, including coagulation, flocculation, sedimentation, and filtration, that results in significant particulate removal.

cross-connection Any actual or potential connection between a drinking (potable) water system and an unapproved water supply or other source of contamination. For example, if a pump moving nonpotable water is hooked into the water system to supply water for the pump seal, a cross-connection or mixing between the two water systems can occur. This mixing may lead to contamination of the drinking water.

Cryptosporidium A disease-causing protozoan widely found in surface water sources. *Cryptosporidium* is spread as a dormant oocyst from human and animal feces to surface water. In its dormant stage, *Cryptosporidium* is housed in a very small, hard-shelled oocyst form that is resistant to chorine and chloramine disinfectants. When water containing these oocysts is ingested, the protozoan causes a severe gastrointestinal disease called cryptosporidiosis.

CT, or CT_{calc} As defined in 40 CFR 141.2, the product of "residual disinfectant concentration" (C) in milligrams per liter determined before or at the first customer and the corresponding "disinfectant contact time" (T) in minutes, i.e., C × T, or CT. If a public water system applies disinfectants at more than one point prior to the first customer, it must determine the CT of each disinfectant sequence before or at the first customer to determine the total percent inactivation or "total inactivation ratio." In determining the total inactivation ratio, the public water system must determine the residual disinfectant concentration of each disinfection sequence and corresponding contact time before any subsequent disinfection application point(s). $CT_{99.9}$ is the CT value required for 99.9 percent (3-log) inactivation of *Giardia lamblia* cysts. $CT_{99.9}$ values for a variety of disinfectants and conditions appear in 40 CFR 141.74(b)(3), Tables 1.1 through 1.6, 2.1, and 3.1. $CT_{calc}/CT_{99.9}$ is the inactivation ratio. The sum of the inactivation ratios, or total inactivation ratio shown as $[(CT_{calc})/(CT_{99.9})]$, is calculated by adding together the inactivation ratio for each disinfection sequence. A total inactivation ratio equal to or greater than 1.0 is assumed to provide a 3-log inactivation of *Giardia lamblia* cysts.

$d_{60\%}$ The diameter of the particles in a granular sample (filter media) for which 60 percent of the total grains are smaller and 40 percent are larger on a weight basis. The $d_{60\%}$ is obtained by passing granular material through sieves with varying dimensions of mesh and weighing the material retained by each sieve.

degasification A process that removes dissolved gases from water. The gases may be removed by either mechanical or chemical treatment or a combination of both.

degradation Chemical or biological breakdown of a complex compound into simpler compounds.

diatomaceous earth (DE) filtration As defined in 40 CFR 141.2, a process resulting in substantial particulate removal that uses a process in which: (1) a "precoat" cake of diatomaceous earth filter media is deposited on a support membrane (septum), and (2) while the water is filtered by passing through the cake on the septum, additional filter media, known as "body feed," is continuously added to the feedwater to maintain the permeability of the filter cake.

direct filtration As defined in 40 CFR 141.2, a series of processes including coagulation and filtration but excluding sedimentation that results in substantial particulate removal.

disinfectant A chemical (commonly chlorine, chloramine, or ozone) or physical process (e.g., application of ultraviolet light) that kills microorganisms such as bacteria, viruses, and protozoa.

disinfectant by-products Chemicals that may form when disinfectants such as chlorine react with plant matter and other naturally occurring materials in the water. These by-products may pose health risks in drinking water.

disinfection by-product precursors Organic or inorganic compounds that react with disinfectants to form disinfection by-products.

distribution system A network of pipes leading from a treatment plant to customers' plumbing systems.

effective range The portion of the design range of an instrument (usually upper 90 percent) in which the instrument has acceptable accuracy.

effective size (E.S.) The diameter of the particles in a granular sample (filter media) for which 10 percent of the total grains are smaller and 90 percent are larger on a weight basis. Effective size is obtained by passing granular material through sieves with varying dimensions of mesh and weighing the material retained by each sieve. The effective size is also approximately the average size of the grains.

effluent Raw, partially treated, or completely treated water or some other liquid that flows from a reservoir, basin, treatment process, or treatment plant.

enhanced coagulation The addition of sufficient coagulant for improved removal of disinfection by-product precursors by conventional filtration treatment.

enhanced softening The improved removal of disinfection by-product precursors by precipitative softening.

enteric Of intestinal origin, especially applied to wastes or bacteria.

entrain To trap bubbles in water either mechanically through turbulence or chemically through a reaction.

epidemic An occurrence of cases of disease in a community or geographic area clearly in excess of the number of cases normally found (or expected) in that population for a particular season or other specific time period. Disease may spread from person to person and/or by the exposure of many persons to a single source, such as a water supply.

exemption State or US Environmental Protection Agency (USEPA) permission for a water system not to meet a certain drinking water standard. An exemption allows a system additional time to obtain financial assistance or make improvements in order to come into compliance with the standard. The system must prove that (1) there are compelling reasons (including economic factors) why it cannot meet USEPA health standards (maximum contaminant levels or treatment techniques); (2) it was in operation on the effective date of the requirement; and (3) the exemption will not create an unreasonable risk to public health. The state must set a schedule under which the water system will comply with the standard for which it received an exemption.

filtration As defined in 40 CFR 141.2, a process for removing particulate matter from water by passing the water through porous media.

finished water Water that has passed through all treatment processes within a water treatment plant such that the water is "finished." This water is ready to be delivered to consumers. Also called product water. See also *source water.*

floc Small particles that have come together (agglomerated) into larger, more settleable particles as a result of the coagulation–flocculation process.

flocculation As defined in 40 CFR 141.2, a process to enhance agglomeration or collection of smaller floc particles into larger, more easily settleable particles through gentle stirring by hydraulic or mechanical means.

fluidization The upward flow of a fluid through a granular bed at sufficient velocity to suspend the grains in the fluid. This process depends on filter media properties, backwash temperature, and backwash water flow rates.

GAC10 Granular activated carbon filter beds with an empty-bed contact time of 10 minutes based on average daily flow and a carbon reactivation frequency of every 180 days.

garnet A group of hard, reddish, glassy, mineral sands made up of silicates of base metals (calcium, magnesium, iron, and manganese) used in the filtration process. Garnet has a higher density than sand.

gastroenteritis An inflammation of the stomach and intestinaal tract resulting in diarrhea, with vomiting and cramps is irritation is excessive. When caused by an infectious agent, it is often associated with fever.

Giardia lamblia A microorganism frequently found in rivers and lakes that is shed during its cyst stage with the feces of humans and animals. When water containing these cysts is ingested, the protozoan causes a severe gastrointestinal disease called giardiasis. People with severely weakened immune systems are likely to have more severe and more persistent symptoms than healthy individuals.

giardiasis Intestinal disease caused by an infestation of *Giardia* flagellates.

grab sample A single sample collected at a particular time and place that represents the composition of the water only at that time and place.

groundwater The water that comes from aquifers (natural reservoirs below the earth's surface).

groundwater under the direct influence (GWUDI) of surface water As defined in 40 CFR 141.2, any water beneath the surface of the ground with significant occurrence of insects or other macroorganisms, algae, or large-diameter pathogens, such as *Giardia lamblia* or *Cryptosporidium,* or with significant and relatively rapid shifts in water characteristics such as turbidity, temperature, conductivity, or pH that closely correlate to climatological or surface water conditions. Direct influence must be determined for individual sources in accordance with criteria established by the state. The state determination of direct influence must be based on site-specific measurements of water quality and/or documentation of well construction characteristics and geology with field evaluation.

haloacetic acids (five) (HAA5) The sum of the concentrations of five haloacetic acids (monochloroacetic acid, dichloroacetic acid, trichloroacetic acid, monobromoacetic acid, and dibromoacetic acid), in milligrams per liter, rounded to two significant figures after addition.

hardness, water A characteristic of water caused primarily by the salts of calcium and magnesium, such as bicarbonate, carbonate, sulfate, chloride, and nitrate. Excessive hardness in water is undesirable because it causes the formation of soap curds, increased use of soap, deposition of scale in boilers, and damage in some industrial processes. Excessive hardness can sometimes cause objectionable tastes in drinking water.

head The vertical distance (in feet [ft]) equal to the pressure (in pounds per square inch [psi]) at a specific point. The pressure head is equal to the pressure in pounds per square inch times 2.31 ft/psi.

head loss A reduction of water pressure in a hydraulic or plumbing system.

health advisory A US Environmental Protection Agency document that provides guidance and information on contaminants that can affect human health and that may occur in drinking water.

humus Organic portion of the soil remaining after prolonged microbial decomposition.

influent water Raw water plus recycle streams.

in-line filtration The addition of chemical coagulants directly to the filter inlet pipe. The chemicals are mixed by the flowing water, and flocculation and sedimentation facilities are eliminated. This pretreatment method is commonly used in pressure filter installations.

inorganic contaminants Mineral-based compounds such as metals, nitrates, and asbestos. These contaminants occur naturally in some water but can also get into water through farming, chemical manufacturing, and other human activities. The US Environmental Protection Agency has set legal limits on 15 inorganic contaminants.

jar test A laboratory procedure that simulates a water treatment plant's coagulation, rapid mix, flocculation, and sedimentation processes. Differing chemical doses, energy of rapid mix, energy of slow mix, and settling time can be examined. A jar test is used to estimate the minimum or optimal coagulant dose required to achieve certain water quality goals. Samples of water to be treated are commonly placed in six jars. Various amounts of a single chemical are added to each jar while holding all other chemicals at a consistent dose and observing the formation of floc, settling of solids, and resulting water quality.

maximum contaminant level (MCL) The highest level of a contaminant that is allowed in drinking water. MCLs are set as close to the maximum contaminant level goal as feasible using the best available treatment technology and taking cost into consideration. MCLs are enforceable standards.

maximum contaminant level goal (MCLG) The level of a contaminant in drinking water below which there is no known or expected risk to health. MCLGs allow for a margin of safety. MCLGs are nonenforceable health goals.

maximum residual disinfectant level (MRDL) A level of a disinfectant added for water treatment that may not be exceeded at the consumer's tap without an unacceptable possibility of adverse health effects. For chlorine and chloramines, a public water system (PWS) is in compliance with the MRDL when the running annual average of monthly averages of samples taken in the distribution system, computed quarterly, is less than or equal to the MRDL. For chlorine dioxide, a PWS is in compliance with the MRDL when daily samples are taken at the entrance to the distribution system and no two consecutive daily samples exceed the MRDL. MRDLs are enforceable in the same manner as maximum contaminant levels under the Safe Drinking Water Act.

maximum residual disinfectant level goal (MRDLG) The maximum level of a disinfectant added for water treatment at which no known or anticipated adverse effect on the health of persons would occur and that allows an adequate margin of safety. MRDLGs are nonenforceable health goals and do not reflect the benefit of the addition of the chemical for control of waterborne microbial contaminants.

microbes (microorganisms) Tiny living organisms that can only be seen with the aid of a microscope. Some microbes can cause acute health problems when consumed. See also *pathogens.*

microbial growth The activity and growth of microorganisms such as bacteria, algae, diatoms, plankton, and fungi.

micrograms per liter (μg/L) One microgram of a substance dissolved in a liter of water. This unit is equal to parts per billion (ppb) because 1 liter of water is equal in weight to 1 billion micrograms.

micron A unit of length equal to one micrometer (μm), one millionth of a meter, or one thousandth of a millimeter. One micron equals 0.00004 of an inch.

microorganisms Living organisms that can be seen individually only with the aid of a microscope.

milligrams per liter (mg/L) A measure of concentration of a dissolved substance. A concentration of 1 mg/L means that 1 milligram of a substance is dissolved in a liter of water. For practical purposes, this unit is equal to parts per million (ppm) because 1 liter of water is equal in weight to 1 million milligrams.

monitoring Testing that water systems must perform to detect and measure contaminants. A water system that does not follow the US Environmental Protection Agency's monitoring methodology or schedule is in violation and may be subject to legal action.

mudball Material that forms in filters and gradually increases in size when not removed by the backwashing process. Mudballs are approximately round in shape and vary in size from pea-sized up to 2 or more inches in diameter.

National Primary Drinking Water Regulations Legally enforceable standards that apply to public water systems. These standards protect drinking water quality by limiting the levels of specific contaminants that can adversely affect public health and that are known or anticipated to occur in public water supplies.

nephelometric A means of measuring turbidity in a sample by using an instrument called a nephelometer. A nephelometer passes light through a sample and the amount of light deflected (usually at a 90-degree angle) is then measured.

nephelometric turbidity unit (ntu) The unit of measure for turbidity. See *nephelometric.*

noncommunity water system (NCWS) As defined in 40 CFR 141.2, a public water system that is not a community water system. A noncommunity water system is either a "transient, noncommunity water system (TWS)" or a "nontransient, noncommunity water system (NTNCWS)."

nontransient noncommunity water system (NTNCWS) As defined in 40 CFR 141.2, a public water system that is not a community water system and that regularly serves at least 25 of the same persons over 6 months per year.

operation and maintenance costs The ongoing, repetitive costs of operating and maintaining a water system, e.g., employee wages and costs for treatment chemicals and periodic equipment repairs.

organic contaminants Carbon-based chemicals, such as solvents and pesticides, that can get into water through runoff from cropland or discharge from factories. The US Environmental Protection Agency has set legal limits on 56 organic contaminants.

organics Carbon-containing compounds that are derived from living organisms.

overflow rate A measurement used in the design of settling tanks and clarifiers in treatment plants that relates the flow to the surface area. It is used by operators to determine if tanks and clarifiers are hydraulically (flow) over- or underloaded. Overflow rate may be expressed as either gallons per day per square foot (gpd/sq ft) or gallons per minute per square foot (gpm/sq ft)—overflow rate (gpd/sq ft) = flow (gpd)/surface area (sq ft).

particle count: The results of a microscopic-scale examination of treated water with a special "particle counter" that classifies suspended particles by number and size.

particulate A very small solid suspended in water that can vary in shape, density, and electrical charge. Colloidal and dispersed particulates are artificially gathered together by the processes of coagulation and flocculation.

pathogens, or pathogenic organisms Microorganisms that can cause disease (such as typhoid, cholera, dysentery) in other organisms or in humans, animals, and plants. They may be bacteria, viruses, or protozoans and are found in sewage, runoff from animal farms or rural areas populated with domestic and/or wild animals, and in water used for swimming. There are many types of organisms that do not cause disease. These organisms are called nonpathogens.

pH An expression of the intensity of the basic or acidic condition of a solution. Mathematically, pH is the negative logarithm (base 10) of the hydrogen ion concentration $[H^+]$—pH = log $(1/H^+)$. pH ranges from 0 to 14; 0 is most acidic, 14 most basic, and 7 neutral. Natural waters usually have a pH between 6.5 and 8.5.

plug flow Water that travels through a basin, pipe, or unit process in such a fashion that the entire mass or volume is discharged at exactly the theoretical detention time of the unit.

polymer A synthetic organic compound with high molecular weight and composed of repeating chemical units (monomers). Polymers may be polyelectrolytes (such as water-soluble flocculants), water-insoluble ion exchange resins, or insoluble uncharged materials (such as those used for plastic or plastic-lined pipe).

pore A very small open space.

precision The capacity for an instrument to measure a process variable and to repeatedly obtain the same result.

primacy Primary enforcement authority for the drinking water program. Under the Safe Drinking Water Act, states, US territories, and Indian tribes that meet certain requirements, including setting regulations that are at least as stringent as the US Environmental Protection Agency's, may apply for, and receive, primary enforcement authority, or primacy.

primary standard A solution used to calibrate an instrument.

public notification An advisory that US Environmental Protection Agency or the state requires a water system to distribute to affected consumers when the system has violated maximum contaminant levels or other regulations. The notice advises consumers what precautions, if any, should be taken to protect their health.

public water system (PWS) As defined in 40 CFR 141.2, a system for the provision of water to the public for human consumption through pipes or, after Aug. 5, 1998, other constructed conveyances, if such system has at least 15 service connections or regularly serves an average of at least 25 individuals daily at least 60 days out of the year. A PWS includes any collection, treatment, storage, and distribution facilities under control of the operator of such system and used primarily in connection with such system, and any collection or pretreatment storage facilities not under such control that are used primarily in connection with such system. A PWS does not include any "special irrigation district." A PWS is either a community water system or a noncommunity water system.

radionuclide An unstable form of a chemical element that radioactively decays, resulting in the emission of nuclear radiation. Prolonged exposure to radionuclides increases the risk of cancer. All radionuclides known to occur in drinking water are currently regulated, except for radon and naturally occurring uranium, both of which were proposed for regulation in October 1999.

raw water Water in its natural state, prior to any treatment for drinking. See *finished water.*

reservoir Any natural or artificial holding area used to store, regulate, or control water.

reverse osmosis The application of pressure to a concentrated solution that causes the passage of a liquid from the concentrated solution to a weaker solution across a semipermeable membrane. The membrane allows the passage of the solvent (water) but not the dissolved solids (solutes). The liquid produced is a demineralized water.

Safe Drinking Water Act (SDWA) Commonly referred to as SDWA. A law passed by the US Congress in 1974.

sample The water that is analyzed for the presence of US Environmental Protection Agency (USEPA)–regulated drinking water contaminants. Depending on the regulation, USEPA requires water systems and states to take samples from source water, from water leaving the treatment facility, or from the taps of selected consumers.

sand Soil particles between 0.05 and 2.0 mm in diameter.

sand filter A filter that uses two grades of sand (coarse and fine) to remove turbidity and particles. A sand filter can serve as a first-stage roughing filter or prefilter in more complex processing systems.

sanitary survey An on-site review of the water sources, facilities, equipment, operations, and maintenance of a public water system for the purpose of evaluating the adequacy of the facilities for producing and distributing safe drinking water.

secondary drinking water standards Nonenforceable federal guidelines regarding cosmetic effects (such as tooth or skin discoloration) or aesthetic effects (such as taste, odor, or color) of drinking water.

secondary standard (for turbidity) Commercially prepared, stabilized, sealed liquid or gel turbidity standards that are used to verify the continued accuracy of a calibrated instrument. The actual value of the secondary standard must be determined by comparing it against a properly prepared and diluted primary standard such as formazin or styrene divinylbenzene polymers. Secondary standards should not be used to calibrate an instrument.

sedimentation As defined in 40 CFR 141.2, a process for removing solids by gravity or separation before filtration.

slow sand filtration As defined in 40 CFR 141.2, a process that involves passage of raw water through a bed of sand at low velocity (generally less than 0.4 meters per hour) to achieve substantial particulate removal by physical and biological mechanisms.

sole source aquifer An aquifer that supplies 50 percent or more of the drinking water for an area.

source water Water in its natural state prior to any treatment for drinking. See *finished water.*

specific ultraviolet absorption at 254 nanometers See *SUVA.*

standard A physical or chemical quantity whose value is known exactly and is used to calibrate or standardize instruments. See also *primary standard* and *secondary standard.*

standardize To compare with a standard. (1) In wet chemistry, to determine the exact strength of a solution by comparing it with a standard of known strength. (2) To set up an instrument or device to read a standard. This allows adjustment of the instrument so that it reads accurately or makes it possible to apply a correction factor to the readings.

state As defined in 40 CFR 141.2, the agency of the state or tribal government that has jurisdiction over public water systems. During any period when a state or tribal government does not have primary enforcement responsibility pursuant to Section 1413 of the Safe Drinking Water Act, the term *state* means the Regional Administrator, US Environmental Protection Agency.

Subpart H system A public water system using surface water or groundwater under the direct influence of surface water as a source that is subject to the requirements of the Surface Water Treatment Rule.

surface water As defined in 40 CFR 141.2, all water which is open to the atmosphere and subject to surface runoff, such as rivers, lakes, and reservoirs.

surfactant Abbreviation for surface-active agent. A chemical that lowers surface tension and increases the "wetting" capabilities of the water when added to water. Reduced surface tension allows water to spread and penetrate fabrics and other substances, enabling them to be washed or cleaned. Soaps and wetting agents are typical surfactants.

suspended solids Solid organic and inorganic particles that are held in suspension by the action of flowing water and are not dissolved.

SUVA (specific ultraviolet absorption at 254 nanometers [nm]) An indicator of the humic content of a water. It is a calculated parameter obtained by dividing a samples ultraviolet absorption at a wavelength of 254 nanometers (UV_{254}) (in m^{-1}) by its concentration of dissolved organic carbon (in milligrams per liter).

total organic carbon (TOC) A measure of the concentration of organic carbon in water, determined by oxidation of the organic matter into carbon dioxide (CO_2).

transient, noncommunity water system As defined in 40 CFR 141.2, a noncommunity water system that does not regularly serve at least 25 of the same persons over 6 months per year. These systems do not have to test or treat their water for contaminants which pose long-term health risks because fewer than 25 of the same people drink the water over a long period. They still must test their water for microbes and several chemicals posing short-term health risk.

treatment technique A required process intended to reduce the level of a contaminant in drinking water.

tube settlers Bundles of small-bore (2 to 3 inches [50 to 75 mm]) tubes installed on an incline as an aid to sedimentation. As water rises in the tubes, settling solids fall to the tube surface. As the sludge (from the settled solids) in the tube gains weight, it moves down the tubes and settles to the bottom of the basin for removal by a conventional sludge collection process. Tube settlers are sometimes installed in sedimentation basins and clarifiers to improve settling of particles.

turbid Having a cloudy or muddy appearance.

turbidimeter A device used to measure the amount of light scattered by suspended particles in a liquid under specified conditions.

turbidity The cloudy appearance of water caused by the presence of suspended and colloidal matter. High levels of turbidity may interfere with proper water treatment and monitoring.

uniformity coefficient A measure of how well a sediment is graded.

USEPA United States Environmental Protection Agency.

variance Permission from a state or the US Environmental Protection Agency (USEPA) allowing a water system to not to meet a certain drinking water standard. The water system must prove that: (1) it cannot meet a maximum contaminant level, even while using the best available treatment method, because of the characteristics of the raw water, and (2) the variance will not create an unreasonable risk to public health. The state or USEPA must review, and allow public comment on, a variance every 3 years. States can also grant variances to water systems that serve small populations and that prove that they are unable to afford the required treatment, an alternative water source, or otherwise comply with the standard.

verification A procedure to verify the calibration of an instrument, such as a turbidimeter.

violation A failure to meet any state or federal drinking water regulation.

virus As defined in 40 CFR 141.2, a virus of fecal origin that is infectious to humans by waterborne transmission.

vulnerability assessment An evaluation of drinking water source quality and its vulnerability to contamination by pathogens and toxic chemicals.

watershed The land area from which water drains into a stream, river, or reservoir.

water supplier A person who owns or operates a public water system.

water supply system The collection, treatment, storage, and distribution of potable water from source to consumer.

wellhead protection area The area surrounding a drinking water well or well field that is protected to prevent contamination of the well(s).

zeta potential The electric potential arising due to the difference in the electrical charge between the dense layer of ions surrounding a particle and the net charge of the bulk of the suspended fluid surrounding the particle. The zeta potential, also known as the electrokinetic potential, is usually measured in millivolts and provides a means of assessing particle destabilization or charge neutralization in coagulation and flocculation procedures.

CPSIA information can be obtained at www.ICGtesting.com
Printed in the USA
BVOW050714040313

314470BV00002B/6/A

9 781583 213858